ELECTRONICS
FOR SCIENTISTS
Practical Essentials
for Modern Research

ELECTRONICS FOR SCIENTISTS

Practical Essentials for Modern Research

Andrew C Haas

New York University, USA

World Scientific

NEW JERSEY · LONDON · SINGAPORE · BEIJING · SHANGHAI · HONG KONG · TAIPEI · CHENNAI · TOKYO

Published by

World Scientific Publishing Co. Pte. Ltd.

5 Toh Tuck Link, Singapore 596224

USA office: 27 Warren Street, Suite 401-402, Hackensack, NJ 07601

UK office: 57 Shelton Street, Covent Garden, London WC2H 9HE

Library of Congress Control Number: 2025032520

British Library Cataloguing-in-Publication Data
A catalogue record for this book is available from the British Library.

ELECTRONICS FOR SCIENTISTS
Practical Essentials For Modern Research

ISBN 9789819821556 (hardcover)
ISBN 9789819821563 (ebook for institutions)
ISBN 9789819821570 (ebook for individuals)

For any available supplementary material, please visit
https://www.worldscientific.com/worldscibooks/10.1142/14535#t=suppl

Desk Editors: Eshak Nabi Akbar Ali/Muhammad Ihsan

Typeset by Stallion Press
Email: enquiries@stallionpress.com

Foreword

Electronics have become central to experimental science, and are only becoming more important with each passing year. They are essential parts of experimental setups, from small quantum computing experiments to the mega particle detectors at the LHC. Most of these electronics are either purchased from professional companies or designed by professional engineers hired by the experiments. But scientists must *understand* them in order to design the setups and experiments in the first place, and integrate them into the experiments optimally. They also need to know some basic electronics to debug, test, and use these fancy setups. Students can often only make good progress doing research on experiments if they have basic electronics knowledge.

Electronics is also a fascinating subject of applied science in its own right. How electrons and EM fields move in materials, semiconductors, cables, air, etc. are important effects for all scientists to have an understanding of.

Yet many University science programs haven't (until now!) dedicated time in their curricula to teach basic electronics. Students are expected, I suppose, to learn such things as "impedance" and how to correctly measure signals using an oscilloscope in their spare time on their own. Most fail to, and then are lost when faced with issues in a modern lab environment. Imagine if every student could program an Arduino to automate some equipment, use test gear to make accurate measurements, debug sources of electronic noise, and make a PCB of simple circuits!

This course is meant to provide an introduction to basic analog and digital electronics, as would be used in modern scientific research experiments, along with their physical underpinnings. It's aimed at the college or graduate level but could be used by students of any age (advanced high-school, in industry, or hobbyists). No prior knowledge beyond introductory physics concepts is assumed. Mostly just high-school level math is needed.

Program Structure

The book contains a mixture of theory material and practical information. Knowing both are essential to being capable with electronics. Each chapter presents this material, and in a classroom setting it could be summarized in a lecture. Related lab exercises are at the end of each chapter. These are also critical to perform in order to understand the book's lessons and gain practical electronics skills.

In a class environment, there would be one lecture each week ($\sim$1 hour) and a lab session each week ($\sim$3 hours). Students can work in pairs, or individually. Access to the lab setup and equipment should be available to the students outside of lab hours as well. A written report would be handed in each week by each student from the lab session, explaining the circuit developed, showing how it operates in real life (with attached pictures, plots, oscilloscope screenshots, etc.), and answering questions from the book. A PCB is also designed, assembled, and tested by each student as a larger project, which is then presented to the class. The whole program is meant to be completed in a single 14-week semester.

AI

The use of AI when learning electronics and performing the lab exercises in this book can be both helpful and harmful. Using it to answer too many questions, and simply doing what it says without understanding why its suggestions are correct, will result in the student learning very little. It would be like a bad teaching assistant that just does all the work for a student in the lab without making sure the student understands and can do it themselves.

On the other hand, a student who is unable to effectively use AI to complement their work and speed their progress is at a real disadvantage in the real world lab. So some middle ground is required. AI should not be outlawed. It should be used to *help a student learn* the material, understand solutions, suggest approaches, and give hints to solve a problem they are stuck on. The word "help" is key. Similarly, a good (human) teaching assistant helps the student learn, without doing everything for them. (I was once taught that the best way to help students in the lab is to "ask them good questions to lead them in the right direction" — a good trick I still use today.) Policing such use of AI is not too hard in a lab setting. An instructor can tell, by talking with the students, whether they are *understanding* what they're doing. When it starts to look like they don't, remind them they're there to learn, not just pass the class!

Equipment Needed

The materials needed for each lab exercise changes somewhat from week to week. But it's best to just have a single inclusive set of equipment and parts that the students can rely on. They can then be creative in their use of the equipment sometimes in ways that instructors didn't envision. Plus then instructors don't have to worry about new setups each week. This list is not exhaustive, and having more options available to students is always a good thing, but here are some basics:

- **Power supplies**: digital, able to make + and − relative to ground (can also use two isolated + supplies), 30V, 5A is plenty. Should have current limiting capabilities, ability to measure voltage and current is a plus.
- **Multimeters**: digital, 3.5 digits, nothing fancy. Good if they can measure capacitance. Have at least two, for simultaneously measuring voltage and current.
- Cables for the multimeters with alligator clips on the end, to clamp on wires that then go into the breadboard.
- Breadboards, medium-sized are fine.
- A bunch of standard through-hole (for breadboards) resistors, capacitors, inductors, diodes, LEDs. These can be bought as "sample sets" and come with an array of values.
- Some standard through-hole transistors (n-channel, p-channel, BJT, Mosfet, and voltage regulators.

- Some standard through-hole (DIP) op-amps.
- Tons of "dupont" breadboard wires in lots of colors.
- **Oscilloscope**: digital, 4 channels, ~100 MHz bandwidth, not too fancy. Too many features are distracting for a beginning student.
- **Oscilloscope probes**: 4 of them, $1 \times 10 \times$ switchable passive probes, ~100 MHz bandwidth.
- **Function generator**: digital, two channels, can make sine, square, pulse, etc. ± 15 V, 50 Ohm output. Good if it can make very short pulses (~20 ns) with fast rise time (~5 ns).
- Cables from BNC to alligator clips for attaching to wires that go into the breadboard.
- A long BNC cable (40+ feet) and some BNC tee's.
- **An Arduino**: Uno is good.
- A servo motor controlled by PWM, i2c, or SPI.
- **An FPGA dev board**: the simpler the better.
- **A computer**: Windows or Linux. Can preinstall Arduino and FPGA and EDA (PCB design) software.
- **Soldering tools**: Iron with adjustable (digital) temperature, several tips, solder of various widths, mesh tip cleaner, no-clean flux, side-cutters, a few different tweezers, wire strippers. Can also have a hot-air rework station in the room for fixing occasional mistakes.

About the Author

 **Andrew Haas** is a professor of physics at New York University. His research is in experimental particle physics, using particle accelerators such as the LHC at CERN to look for new fundamental particles that make up our Universe. He is part of the ATLAS team that discovered the Higgs boson in 2012, and now also co-leads a new experiment at CERN looking for particles with small electric charge. His work focuses on the electronics and software used to record and understand the data from these particle detectors. For years, he has taught electronics to undergraduate science majors and graduate students. He also designs and makes open source electronics hardware, firmware, and software for fun and education.

Contents

Chapter 1

DC Circuits

Let's start with a very simple circuit. A classic flashlight has four components: a voltage source, a switch, and a lightbulb, all connected by wires. (More specifically, the wires connect them in *series*, but we'll discuss what that means later.) (To be a *circuit* means the wires must connect components in a closed path.)

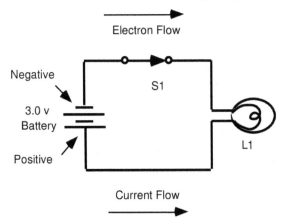

Voltage

Electronics is all about the study and control of electric charge. We assume you know that charge can be positive or negative, and that opposite charges attract, whereas like charges repel. Charge generates an electric field, and a moving charge also makes a magnetic field, which then both can exert forces on other charges. These basic principles, along with the way charges interact with materials (insulators, conductors, etc.), make all of electronics possible.

Charge is measured in Coulombs (C). One electron has been measured to be about $1.602176634 \times 10^{-19}$ C, sometimes called an amount of charge e. Protons have a charge exactly opposite to that of electrons, which makes atoms neutral when they have an equal number of protons and electrons. Note that 1 C of negative charge is the reciprocal of e, or about 6.241509×10^{18} electrons. That is a big number, which means we can usually think of charge as continuous, and not worry about the fact that it is actually carried by indivisible particles.

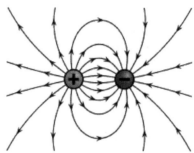

The voltage source, usually made from a battery or two, has the ability to separate positive and negative electric charge. It takes energy to do this, since positive and negative charges attract, and work (force × distance) must be done on the charges. Practically speaking, electrons are easily moved whereas protons are far heavier, stuck in nuclei, crystallized into solid lattices, etc. So what a voltage source usually does is move electrons in some direction, creating a net negative charge there, and leaving behind an area with a lack of electrons, so a net positive charge there. You can think of a battery as a pump, which uses its internal chemical potential energy to push negative charges towards its negative end, leaving net positive charge at its positive end.

Voltage is the potential energy difference that each Coulomb of charge has when the positive charge is separated from the negative charge, compared to when the charge is reunited. For instance, if the battery did 1 Joule (J) of work and separated 1 C of positive charge from 1 C of negative charge, that positive charge would have 1 J/C of potential energy, compared to 0 Joule/Coulomb of potential energy when the charges were together (assuming a perfectly efficient transfer of energy). We got tired of saying Joule/Coulomb, so it was named the Volt (V). Thus there would be a voltage of 1 V where the positive charge is, compared to 0 V where the negative charge is.

Mathematically, voltage is the path integral of moving a charge from one point (a) to another (b) in an electric field (E):

$$V_{ab} \equiv V_a - V_b = - \int_b^a \vec{E} \cdot \vec{dr} \qquad (1)$$

Thanks to the conservation of energy, the path taken from a to b doesn't matter. The electric force is *conservative*, which means it comes from a potential energy. This allows us to meaningfully assign voltages to points.

Always remember, voltage is relative! It means nothing to say that a place in a circuit is at 5 V. You mean it has 5 V compared to some other place in the circuit which you have defined to be at 0 V. This is just like it is useless to say you are driving at 50 km/h. You mean 50 km/h relative to some other thing, like the ground just below you, which you have defined to be moving at 0 km/h. Similarly, we usually call this reference voltage of 0 V *ground*. In practice, we often say that a place in a circuit is at some particular voltage, and it is implied that it has that voltage relative to ground. By the way, there's no problem with having negative voltages, like −5 V, relative to 0 V ground.

Note that the voltage source here is a direct-current (DC) voltage source. That means the voltage at the positive end of the battery (relative to ground, which is at the negative end of the battery!) is constant as a function of time. If it is 1.5 V, then it always has been 1.5 V and always will be 1.5 V. Of course real batteries do lower their voltage over time due to chemical changes within. But over a short period of time, say 10 seconds, the voltage will be very close to constant (assuming the bulb is not too bright). We will later encounter alternating-current (AC) voltage sources (such as that coming from your wall) that switch from positive to negative and back to positive again at some frequency. But for now we'll just consider DC voltage.

Wires

In some materials, such as metals, electrons can easily move when a force is exerted on them. We call such materials *conductors* of electricity, and our wires will be made of them. Materials that don't have free charges, like plastic, wood, air, etc., are called *insulators*. Since there's an excess of positive charge at the positive end of the batteries, electrons in the wire near it will be attracted towards it, resulting in negative charge from the wire near the positive end of the battery and a net positive charge on the other end of the wire, from a deficit of electrons relative to protons. The excess negative charge from the wire will try to cancel some of the excess positive charge from the battery at the positive end of the battery. But the battery will continue to do its job of creating excess positive charge, and thus a constant positive voltage, at its positive end. This process will continue until there is a single, stable, constant voltage at every point along the wire connected to the positive end of the battery. The proof of this is simple: if there were a voltage difference between any two points on the wire, there must be an electric field pointing from high voltage to low voltage (see Equation (1)), but charges would then move until the electric field is zero everywhere. This is identical to the way water molecules move freely within a liquid and will do so within a glass of water until they are all at the same gravitational potential energy, and the surface of the liquid is flat after some time (assuming there is some friction!), no matter what shape it started with!

If our switch is open, then nothing else happens. There is a wire connected to the top of the battery with a voltage of 1.5 V everywhere, and a wire connected to the bottom of the battery with a voltage of 0 V everywhere (since we defined our 0 V ground point to be the bottom of the battery). Everywhere in the light bulb and the wires connected to it will also be at 0 V. In the air between the two terminals of the switch, there is a voltage difference of 1.5 V, so there is an electric field pointing from the positive side of the switch (connected to the positive end of the battery) to the negative side of the switch. If there were electric charges in that area that could move freely, they would, but alas there are none, so nothing happens. (Assuming there is air around the switch, there are electrons in air, of course, but almost all of them are bound to the protons in air in neutral atoms and are not free to move. There are a few ions, atoms with a net charge, present that can drift in the electric field. But they are rare, and air is a very good insulator — at normal electric field values at least. If the electric field exceeds about 30 kV/cm, the air molecules will start to have electrons ripped away from them, and then there are free charges to conduct electricity — you get a spark! We will generally try to avoid such extreme situations, so we will keep our switch terminals 1 cm apart, and our voltage to much less than 30 kV.)

Note that even though the voltage distribution here is very simple, the charge distribution (where there is excess positive or negative charge, and how much) is quite complicated. It would depend in detail on the geometry of the setup, how long the wires were, whether they were straight or bendy, etc. (Imagine bending the wire connected to the negative terminal such that it went close to the positive terminal — there would then be some region in the wire there with a net negative charge, because it is attracted to the excess positive charge at the battery terminal, leaving a net positive charge elsewhere in the wire — complicated! But the wire still has a voltage of 0 V everywhere!) Without an extremely detailed numerical simulation, all we can really say is there's an excess of positive charge near the positive switch terminal, and an excess of negative charge at the negative switch terminal. That's the only way we could have an electric field pointing from the positive switch terminal to the negative terminal. Almost always, we just won't care about the charge distribution in a circuit. All functioning of the circuit can be derived from the voltages alone, and thanks to the path independence of Equation 1, they are far simpler to analyze. Imagine how difficult electronics would be if we had to worry about the exact shape and length and angles of all our wires!

Current and Resistance

Ok, let's finally turn on the flashlight! We close the switch. What happens? Electrons start flowing through the closed switch, since they are repelled by the excess negative charge on the side of the switch near the bulb and attracted by the excess positive charge on the side of the switch near the positive battery terminal. Remember that electrons have negative charge! You could also say that there is an electric field across the switch pointing from high voltage to low voltage, and electrons are accelerated by this field, towards higher voltage. This would continue until electrons had moved such that there is no longer a voltage difference, and thus no longer an electric field, from one side of the switch to the other. (We discussed this same idea before for electrons in a wire, moving so as to eliminate any voltage differences along a wire.)

But if the wire to the right of the closed switch is now also at 1.5 V, we have a voltage difference (and thus an electric field) across the light bulb! So, electrons will start to move from low towards high voltage within the bulb. How long does it take for electrons to start flowing through the bulb after you close the switch? You only have to wait for the new electric field (or voltage) to arrive at the bulb, which travels at the speed of light (roughly). All the electrons in the circuit start flowing nearly simultaneously. You don't have to wait for electrons to flow from the switch to the bulb in the wire. That would take a long time! We'll calculate about how long below.

Something strange happens to the electrons as they move through the light bulb. They lose energy along the way and transfer their energy into heat (which then illuminates the light bulb)! The process is a bit complicated and quantum mechanical in detail, but it is analogous to friction. The electrons are "bumping" into other atoms along their journey through the bulb and transferring their energy to these other atoms and heating them up. The result is that the electrons (on average) had 1.5 V of energy/charge (J/C) on one side of the bulb, but then exited with 0 V on the other side of the bulb. The more energy each electron loses on average as it moves through the bulb, the fewer electrons we need to move through the bulb each second in order to have them lose an average of 1.5 J/C each second. In other words, the larger the "resistance" of the bulb, the smaller the required flow rate of electrons through the bulb.

We call the rate of charge flowing through a wire "current", usually written with the letter "I". (I'm not sure why "I" is used, but I guess "C" was already taken for charge or Coulomb, so may as well!) The units of current are Coulomb/second (C/s), also known as Amperes (A). So if there is one Coulomb, or about 6.241509×10^{18} electrons, passing through the light bulb each second, there is a current of 1 A flowing through the bulb. Current is defined such that current flows from high voltage towards low voltage. This means that electrons, since they are negatively charged, are actually flowing opposite to the direction of current! But that's OK. We always just talk about the flow of current, and (*almost* always) just forget about the fact that electrons are actually flowing the other way. Note that since charge is conserved, and because this circuit has V and I which are constant in time, the current is the same at all points in the circuit! Otherwise charge would continuously build up at some point in the circuit. So there is also 1.5 A flowing through each point in all the wires, and through the battery.

We can now be more precise and quantitative about the relation between charge flow through the bulb (I) and the resistance (R) of the bulb. What we said above is that I is inversely proportional to R, so I is proportional to $1/R$. But if the voltage were larger, we would also need more current to flow, because each electron loses a fixed amount of energy in the bulb on average, but we need to lose more energy/charge going from one side of the bulb to the other. So I is also proportional to V. Putting it together, $I = V/R$. This is Ohm's law (sometimes written $V = IR$) and provides the quantitative definition of resistance. R is measured in units of V/I, or $(J/C)/(C/s)$, or $J * s/C^2$, which we also call Ohms (Ω). So if there is a voltage of 1.5 V from one side of the bulb to the other, and a current of 1.5 A is flowing through the bulb, the bulb has a resistance of $1\,\Omega$. You can also use the formula in other ways, to e.g. solve for V if you know I and R, or I if you know V and R. But in all cases, be careful what V, I, and R mean. V is the voltage difference between two points, R is the total resistance between those two points, and I is the current flowing from one point to the other (or equivalently, through either point — there is the same current flowing through both points).

Warning! Ohm's "law" only works for *some* things. In general, any simple material, like a metal wire, or a piece of carbon, will obey Ohm's law. An old fashioned light bulb has a medium-resistance metal wire suspended in vacuum (or an inert gas) inside glass, so Ohm's law is good there. Wires made out of metal will have some (small) resistance (which we approximated as zero in our discussion above), and Ohm's law is good there too. For *many* other things, like diodes or transistors (discussed later), Ohm's law simply does not apply.

And even for things that do obey Ohm's law, their resistance is never truly a constant. In reality it will depend on things like temperature, pressure, humidity, magnetic field, time (age), etc. "Resistors" are usually made by manufacturers to be as stable as possible with respect to these external factors. But for any particular resistor you would have to read the datasheet for the part to understand its dependencies.

We can now estimate how fast the current is flowing in the wire, given it's 1.5 A of current, or about $6.241509 \times 10^{18} \times 1.5$ electrons per second. There are about 6.022×10^{23} free electrons (Avogadro's number) in each mole of copper in the wire, one electron per copper atom. One mole of copper weighs about 63.5 grams, which given the density of copper at room pressure is about $7110 \, mm^3$, or a $1 \, mm^2$ thick wire 7.11 m long. But we only have $6.241509 \times 10^{18} \times 1.5$ electrons passing per second, not 6.022×10^{23}, so we need only the electrons in $7110 \, mm \times 6.241509 \times 10^{18} \times 1.5 / 6.022 \times 10^{23} = 0.1105 \, mm$ of wire to pass each second. Our current only flows at about 0.1 mm/s! Fortunately, the changes in the electric field propagate at nearly the speed of light, or $3 \times 10^{11} \, mm/s$, a few trillion times faster.

Power

How much heat and light are generated in the light bulb each second? Or, equivalently, how much energy is being taken from the battery each second? Energy/time is called power, so we are asking about the power used from the battery and given out from the bulb. Well, the change in voltage (e.g. from one side of the bulb to the other) is the change in energy per charge, J/C. And the current is the number of charges per second, C/s. So multiplying the two gives the energy change per second, J/s, which is power: $P = V * I$. So there is $1.5 \, V \times 1.5 \, A = 2.25 \, J/s$ of power being used in the bulb. We get tired of saying J/s so call it Watts (W).

If Ohm's law can be applied to the situation at hand, then $V = IR$, and you can use that information to derive that $P = I^2 R$ and V^2/R. These last two equations tell important lessons to remember. If you need to run twice the current through some resistance, or have twice the voltage across some resistance, the power needed (and heat generated each second) is not twice as large, it's *four* times as large!

Calculating the power used is a *very* important part of designing any circuit, even simple tests. Power usually goes into making heat, unfortunately. If you're not careful, your circuit (or some piece of it) will melt or vaporize! Or less dramatically, some component will just get too hot and fail.

Fields?!

Let's take a moment to consider something people usually don't: what is the electric field created by this circuit? We know what it is inside the battery, because E is the slope (gradient) of voltage, which goes from 1.5 V at the top to 0 V at the bottom, so it's (1.5 V–0 V)/(battery length). A similar calculation gives the E field in the light bulb (resistor). In the wires, there is no E field when the switch is off, since the voltage is constant in the wires. But when the switch is on, the voltage is still constant in the wires, but current is flowing in the wires — so there *must* be an electric field in the wires which is pushing charge (electrons) through the wires! What about outside the circuit in the air? There's an electric field there too! In the air near the positive end of the battery, there's certainly an electric field pointing outward from the accumulation of positive charge there. We just don't usually care about the electric field in the air around a circuit because there's (almost) no free charges there which can move or react to the electric field in any way.

There's also a magnetic field outside the circuit. This is easier to calculate because it's just given by the Biot–Savart law and the right hand rule, which says the magnetic field B (or more generally, H) in a circle around a wire is proportional to the amount of current flowing through the circle, and inversely proportional to the radius of the circle. In total, the electric and magnetic fields around our circuit, when it is on, look something like this:

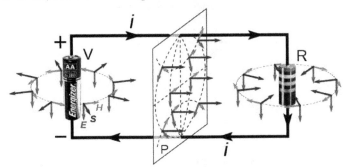

The Poynting vector in electromagnetism, $S = E \times H$, gives the flow of energy being carried by electromagnetic fields. For instance, for a ray of light, which has perpendicular electric and magnetic fields, it gives the amount of energy, and the direction of energy, being transmitted by that light. In this case for our circuit, it says that the energy is mostly flowing through the space outside the wires! Energy flows out from the *side* of the battery, in the space around and between the wires, and then in towards the light bulb (resistor) from the side! This is actually an important lesson — energy does not flow in the wires of the circuit, it is *guided* by the wires in the circuit, because the wires are where charges can move, and it is those moving charges which create magnetic fields perpendicular to electric fields.

The Voltage Divider

Now let's make our circuit just a little more complicated — we'll add a second resistor (or a second lightbulb). There are two topologically different ways to do so. The first way would be to add our second resistor in "series" with the old one, meaning that all current that goes through the first resistor (R_1) then goes into the new one (R_2). (seen below on the left) In that case, the energy lost by a charge must be the sum of the energy lost in R_1 plus the energy lost in R_2. So the total resistance, $R = R_1 + R_2$.

The other option is to add the second resistor in "parallel", meaning that current *splits* and then some goes through R_1 and the rest goes through R_2. (seen below on the right) What is the total resistance of both resistors in this case? The way to think of it is that the total "conductance", $1/R$, is the conductance of R_1, which is $1/R_1$, plus the conductance of R_2, which is $1/R_2$.

$$1/R = 1/R_1 + 1/R_2 -> R = 1/(1/R_1 + 1/R_2) = R_1 * R_2/(R_1 + R_2)$$

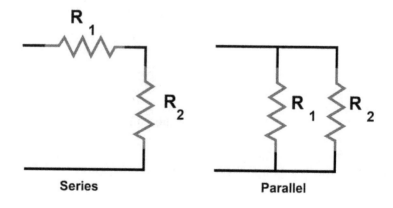

Series   **Parallel**

The case of two resistors in series, makes perhaps the most basic and important circuit in all of electronics, the "voltage divider":

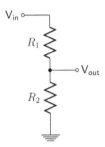

(The symbol at the bottom of R_2 means that is a connection to Ground, i.e. V_{in} and V_{out} are voltages relative to that point.)

The output voltage, V_{out}, depends on the input voltage, V_{in}, but also on the values of R_1 and R_2. We can use Ohm's law (twice) to calculate V_{out}. Think for a moment yourself how you would do so.

The current through R_1 and R_2 is given by $I = V/R = V_{in}/(R_1 + R_2)$. The voltage across R_2 is V_{out}, what we're after, and it is given by $V = IR = V_{in} * R_2/(R_1 + R_2)$. So the lesson to remember here is that the fraction of V in that goes to V_{out} is given by $R_2/(R_1 + R_2)$. This makes sense. If R_2 is really small, then V_{out} must be really close to Ground. But if R_2 is large, V_{out} is mostly just connected to V_{in} through a resistor, so is close to V_{in}. You can also think about taking the limits as R_2 goes to 0 or infinity.

Output Resistance and Load

This voltage divider is our first, very simple, example of a circuit *stage*. Any complicated electronic device, such as a guitar amplifier, is going to have multiple stages. Each stage takes in some input(s) and produces some output(s). For instance, the first stage might be a pre-amplifier which takes the input signal from the guitar cable and amplifies it to a given standard voltage range (say -1 to $1\,V$ peak-to-peak) and then outputs that signal to the next stage. That next stage might have some filter which boosts high frequencies and removes low frequencies and outputs to the next stage. The final stage might be a power amplifier which takes the filtered signal and provides the (large) current to move the speaker. This is the only way we humans can ever hope to make a complicated circuit work!

We design each stage one-at-a-time, and test each one one-at-a-time, with known inputs, and measure the outputs. Then once we know each stage works independently, we can combine them together into a complete circuit. Here, the voltage divider stage simply takes an input signal (V_{in}) and produces an output signal (V_{out}), where V_{out} is smaller than V_{in}.

But in order for this approach to work, we need the output of each stage to not be heavily affected by the input of the stage following it. Let's take a look at the voltage divider again and see what happens as we put different *loads* on it. A load is just a resistance attached to the stage output, V_{out} in this example, and going to Ground (usually). Let's put a load of 450 Ω on this voltage divider made from two 100 Ω resistors, as shown in the following figure.

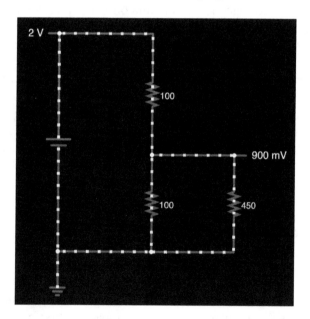

In theory, we said the voltage at V_{out} should be $V_{in} * R_2/(R_1 + R_2) = 2V * 100/200 = 1\,V$. This would be correct if there was no load (infinite resistance to ground) on the output. But after we attach the load of 450 Ω, V_{out} has dropped to 0.9 V! (Why?) So in general, loads, or following stages, will affect the outputs of earlier stages, usually by lowering V_{out}. Here, a 450 Ω load only reduced V_{out} by 10% (0.1 V/1 V). But what if we attached a load of 50 Ω? Then the output drops by 50% to 0.5 V! Big effects like this would ruin our ability to accurately combine multiple circuit stages. How do we make sure the following stage won't affect the previous output too much?

Well let's see if we can figure out quantitatively what's going on. With the $450\,\Omega$ load, there is $I = V/R = 900\,\text{mV}/450\,\Omega = 2\,\text{mA}$ flowing through the load. So for a load current of $2\,\text{mA}$, the V_{out} load voltage decreased by $0.1\,\text{V}$. It's like the output has a resistance of $R = V/I = 0.1\,\text{V}/0.002\,\text{A} = 50\,\Omega$. Let's check that the output resistance stays the same for the $50\,\Omega$ load. Well, the voltage decreased by $0.5\,\text{V}$ and the load current is $I = V/R = 0.5\,\text{V}/50\,\Omega = 0.01\,\text{A}$. So it's again like the output has a resistance of $R = V/I = 0.5\,\text{V}/0.01\,\text{A} = 50\,\Omega$. This voltage divider stage, made with two $100\,\Omega$ resistors, seems to act like it has an output resistance of $50\,\Omega$.

In general we can always find the output resistance by changing the output voltage (by varying the load resistance) and seeing what happens to the output current. Then $R = \Delta V/\Delta I$. For the voltage divider, we found that the output resistance was $50\,\Omega$, from two $100\,\Omega$ resistors. This happens to be the resistance of two $100\,\Omega$ resistors in parallel, $R = (R_1 * R_2)/(R_1 + R_2) = 100*100/200$. It's not an accident — this is the output resistance for any voltage divider. In general, to find the output resistance for a circuit stage, you can also just "look back" towards the input from the output, with the voltage input shorted to ground (to account for the fact that V_{in} itself is assumed to be a perfect voltage source with no resistance). For the voltage divider, after connecting the input V_{in} to ground, looking from V_{out} back towards ground, there are two parallel paths from V_{out} to ground, one through R_1 and another through R_2, so the total resistance is just that of R_1 and R_2 in parallel.

As long as the resistance of the load on a stage is much greater than the output resistance of the preceding stage, the load's effect on the preceding stage will be small. For example, we had $50\,\Omega$ output resistance of the voltage divider, so as long as the load is more than about $10\times$ larger resistance ($500\,\Omega$ or above), the impact on the voltage produced by the preceding stage (V_{out}) will be small, less than a 10% effect. Input and output resistance determine which way "information" flows through a circuit, i.e. which side is the input vs. the output. We want changes at our inputs to affect the results at the outputs, not vice versa!

Also be wary of "overloading" a circuit when measuring it. Measuring the circuit means you have to interact with it, using some resistance, similarly to how you must affect a system in quantum mechanics in order to make a measurement! Hopefully this resistance is high, like $10\,\text{M}\Omega$ for a digital multimeter or oscilloscope probe (in $10\times$ mode!), and it is more than $10\times$ the output resistance of the part of the circuit you are testing. But we'll see cases later on (at higher frequencies) when things are not so easy.

Input Resistance

What if we wanted to follow this voltage divider stage by another stage, not just a simple resistor load? We know the input resistance of a simple resistor load (it's just the resistance of the resistor!), but what is the input resistance of some more complicated circuit stage? We need to be able to find it in order to check that it's going to at least $10\times$ the output resistance of the preceding stage. Well you just apply a voltage to the input, and see how much current flows into the input. The ratio $R = V/I$ gives the input resistance. It's best to do this with no load attached to the output V_{out} of the stage. For our voltage divider, for instance, if we apply a voltage of $2\,\text{V}$ to the input, a current of $0.01\,\text{A}$ flows into the input, since $I = V/R = 2\,\text{V}/200\,\Omega$, where $200\,\Omega$ is the total resistance of R_1 and R_2 in series. So our input resistance is $R = V/I = 2\,\text{V}/0.01\,\text{A} = 200\,\Omega$. The input resistance is just $R_1 + R_2$ for any voltage divider in general.

Power and Matched Impedance

What about the power used by our voltage divider? We might be concerned about this if, for instance, we don't want our resistors to burn up! Well $P = VI = V^2/R$, so for any given input voltage V_{in}, the power used will be inversely proportional to the input resistance. So if we want to use little power, we need to use big resistors. But there's a downside! Those big resistors will also lead to a larger output resistance! There's no free lunch in electronics. In order to have a lower output resistance from our voltage divider (or any circuit stage in general), we have to pay for it by using more power. We will see a very clever way to mostly avoid this issue later on (using active components).

Lastly, there are some special cases where it may be best to break this $>10 \times$ input/output resistance rule of thumb. For instance, what if the goal is to transfer the most power to the output load, for a given input voltage V_{in}. Well, the power in the load is $P = VI$. We can increase I by decreasing load resistance, since $I = V/R$. But wait — the voltage V across the load will also get smaller if we decrease the load resistance! In the limit that R goes to 0, the voltage across the load also goes to 0 (it is shorted to ground!), so $P = VI = 0$. The power in the load also clearly goes to 0 in the limit of R goes to infinity, because then I goes to 0. So the most power in the load is for some value of R between zero and infinity. It turns out that the most power in the load is when the load resistance is equal to the output resistance of the voltage divider! (See if you can work this out for yourself based on the formulae for the voltage and the current through the load as a function of load resistance.) This is a very general result in physics. For instance, in optics, the most power (light) is transmitted through an interface between two materials (and the least amount of light is reflected from the interface) when the index of refraction of the second material is equal to that of the first material. We say that they have the same *impedance*.

Lab: DC Circuits Lab

Setup

Before you start, take the time to *clean and organize* your workspace. That means wiping down your benchtop surface with a damp paper towel. A little isopropyl alcohol is great for removing oils too, and the 99% stuff is non-conductive. In addition to removing dust and grime that could interfere with circuit connections, it gives a level of respect to the process on which you're about to embark. Be proud of your circuit! When you take a picture of it, you want it and its surroundings looking nice.

Then organize your supplies and tools. Think ahead of time where to put your meters and instruments, your breadboard, and your supplies. Wires and components should be clearly separated from each other and easy to find. Think about where you are going to stand or sit, and whether you (and your partner if you are a team) and the lab instructor will be able to easily see instrument displays, reach the circuit being worked on, and be *comfortable*.

I know this all sounds silly, but it's absolutely essential, especially when later working on complicated circuits and projects, so start the habit from day one. It will make working on your projects a pleasure, rather than confusing, uncomfortable, and stressful.

Breadboards

We'll use breadboards for most circuits in this class, since they make it easy to quickly *prototype* circuits. You can make and change connections between components easily.

A nice introduction to breadboards is here: https://learn.spark-fun.com/tutorials/how-to-use-a-breadboard.

There are limitations. Breadboards quickly become a total mess with wires going everywhere, as soon as you have more than a few components. And later we'll understand how capacitance (between pins) and inductance (of wires) limits the speed of circuits on breadboards. But they're great for testing simple, slow, circuits.

Even for a very simple circuit, it's essential to practice good breadboard techniques:

1. Use the right color of wire for a given part of the circuit. Red is for positive voltage input power. I like blue for negative voltage input power. Green or black is for ground. All the other colors can then be used for other intermediate connections. Try to not reuse a color for two different nearby signals. For instance, don't have a yellow wire coming into a resistor and then another yellow wire coming out the other side of the resistor. At some point some day you're going to be tempted to be lazy and use that black wire sitting right there next to you for an intermediate signal, or even (gasp) a positive voltage — don't do it — you'll regret it later!

2. Layout of the circuit on the breadboard should be like a good schematic drawing. Power flows from high voltage down to low voltage. So you should have positive voltage coming into the top of your circuit on the breadboard, then going through components, and ending up at ground at the bottom of the circuit. (If there are negative voltages, they should be lower than ground.) *Causality* should flow from the left of the circuit to the right, from inputs on the left to outputs on the right. For example, a voltage divider which creates an intermediate voltage as an output may then go to

the input of a following circuit stage (like a load, or a filter, or an amplifier, etc.). The voltage divider should be on the left, showing its output wire heading out to the right, going into the input of the following stage. These standards make it easy to understand for someone else who looks at your circuit, and will help you avoid errors in your layout connections.

Multimeter

To measure electricity, the number one tool is the multimeter. (Unfortunately, electricity is mostly invisible!) Multimeters are slow, so you can only use them for voltages and currents which are not changing too quickly. For measuring rapidly varying electrical signals you need an *oscilloscope*. A good intro to the multimeter is here: https://www.youtube.com/watch?v=52w3xeXrMU8&t=2s.

Probably the most common confusion with multimeters for beginners is how to measure current. For almost all other measurements, like measuring voltage, your multimeter is in *parallel* with the rest of the circuit. It then has very high resistance, about $10\,\text{M}\Omega$ usually, so it's just spying on the circuit a little bit from the outside. To measure current, the multimeter has to go in *series* with the circuit at the point where you want to measure the current. This means you have to break the circuit temporarily, at the point where you want to measure the current, then make that current flow *through* your multimeter in order to complete the circuit and measure the current. In this mode, the multimeter has very low impedance (around $0.1\,\Omega$ usually) so it doesn't influence the behavior of the rest of the circuit significantly.

Simulation

Computers make it easy these days to simulate circuits, even very complicated ones. For any circuit you want to build, it's a good idea to first simulate the circuit. If it doesn't work in simulation, it's certainly not going to work on the breadboard! If it does work in simulation, but not on the breadboard, you can try to figure out why it doesn't work as expected, usually by measuring voltages at various points from left to right in the circuit until you find discrepancies. You can also explore more minor differences between the simulated results and the measurements of the real circuit, which may come from *non-ideal* behavior of components in your circuit. (For example, wires are not perfect conductors, resistance changes with temperature, everything has capacitance and inductance, and there is noise everywhere!)

There are many circuit simulators out there. For this class, we use *Falstad* https://falstad.com/circuit/, which is simple, needs no installation, has a good array of example circuits, and can simulate a sufficiently broad range of circuits.

Safety

As you know, electricity can be dangerous. But as with anything, it can be handled safely as long as we understand it. The basic thing you want to avoid is too much electrical current flowing through you. This will happen if one part of your body touches a conductor with one voltage and another part touches a conductor with another voltage. There is then a voltage difference across your body (or part of your body) and since $I = V/R$, current will flow. The amount of current depends on the voltage difference, as well as the resistance of your body in the circuit. This resistance mostly depends on the points of contact between you and the circuit, since your body itself internally is a pretty good conductor, basically that of salt water. Your skin provides pretty good resistance, around $1\,\text{M}\Omega$, depending on how dry it is and how large the contact area is. Go ahead and measure your resistance with the multimeter!

In this class we will only work with "low" voltages, meaning $\sim 15\,\text{V}$ or less (relative to ground). You could have $+$ and $-15\,\text{V}$, so a voltage differential of up to $30\,\text{V}$. As long as your fingers are reasonably dry and no other part of your body is making a good connection to ground (wear shoes and don't stand in water!), it's unlikely you would feel a shock from such voltages. But better yet, just don't touch your circuit while it's powered on!

For more info here's a good source: https://www.allaboutcircuits.com/textbook/direct-current/chpt-3/shock-current-path/.

Exercises

1. Measure the current through a resistor as a function of voltage across the resistor (including negative voltage). Don't exceed ±15 V. Calculate the power used at each voltage, and make sure not to exceed 1/8 W in the component (or it may burn up!). Compare your measurements to theory from Ohm's law. Try a few different resistances.

2. Do the same but for a 914B diode. Be careful not to pass more than 100 mA through the diode (or it may burn up!). You can limit the current using resistance in series with the diode. Be careful to then measure current with respect to the voltage across the diode.

3. Make a voltage divider and measure the voltage at the output, compared to the input voltage, for a few different divider ratios. Compare your measurements to theory and simulation.

4. Attach various loads (resistances to ground) to a voltage divider output, and compare to theory and simulation. How much current goes through the load? How much power is used in the load? What's the relationship between power given to the load vs the load resistance? How does this relate to the output impedance of the voltage divider?

Chapter 2

Capacitance and Filters

Up to now, our circuits have been steady as a function of time. We studied what the flashlight was like while it was "off", and what was happening while it was "on". In both states, everything was constant. The voltages and currents (if there were any) were unchanging while it remained in the same state. That was a good place to start (and fairly complicated already!), but nearly every useful circuit has voltages and currents that are always changing, or at least quite often. For instance, an audio amplifier is taking in a changing voltage as a function of time (the input signal) and outputting a larger changing voltage and current (the output signal making the speaker move). Or a sensor is deciding every 1 us (at 1 MHz) whether to set the voltage on a wire to 3.3 V or 0 V, to represent a binary 1 or 0, respectively, on a digital signal wire attached to a microprocessor input. Even our flashlight wouldn't be so useful if you couldn't turn it from off to on (and vice versa), which must change the current from 0 A to 1.5 A!

So OK, what happens when we turn on the flashlight? We sort of addressed this before — we said that current starts flowing through the bulb very quickly, as soon as the electric field change propagates from the switch to the bulb, a time not much more than the distance from the switch to the bulb divided by the speed of light, much faster than the time it takes electrons to flow from the bulb to the battery. But when the new electric field gets to the bulb, does the voltage across the bulb instantly go from 0 to 1.5 V? Of course not! The voltage across the bulb is going to be caused by a buildup of positive charge at the positive voltage side of the bulb, and a corresponding buildup of negative charge at the other side. It is going to take *time* to build up that excess charge at those places. The excess charge must move there (a current must flow), and that takes time. Note — the excess charge does *not* have to travel from the switch to the bulb — the charges in the wire near the bulb just move a tiny bit towards the bulb. We'll calculate soon just how many charges are needed and thus how far they have to travel, given the density of free charges in the wire, but — spoiler alert — it's not far!

Capacitance

How much charge, Q, do we have to build up on each side of the bulb in order to create a voltage, V, across the bulb? Let's call this ratio, Q/V, *capacitance*, C. So $C = Q/V$. Would it always take the same amount of charge buildup to create a given voltage difference from one side of an object to another? In other words, do all objects have the same capacitance, or does it depend on the details of the object somehow?

Let's study a particular object and calculate its capacitance. Imagine two parallel metal plates, each of area A, connected to opposite ends of a battery. Charge will build up on each plate until the voltage difference between the plates is equal to the voltage of the battery, V, at which point the amount of charge, Q, is stable. So we're back to the original question — how big is Q, for a given V?

The electric field, E, between the plates should be fairly uniform, as long as the plates are large compared to the distance between them, d. Then the energy needed to move a positive charge from the negative plate to the positive plate is $E * d$, because work = force * distance, and the electric force on a charge is $F = E * q$, or just E for a unit charge ($q = 1$). Voltage is defined as energy per unit charge (a V is a J/C). So the voltage difference between the plates must be $V = E * d$.

We can also use Gauss's law to directly calculate the electric field, and find that $E = Q/(A * \varepsilon_0)$, where ε_0 is the vacuum permittivity. (Gauss's law says that the integral through a closed surface of $E * dS$ equals the integral of the charge within that surface divided by ε_0, so $E * A = Q/\varepsilon_0$.) If our plates have something other than vacuum (emptiness) between them, then this constant can be larger, so would be multiplied by some additional constant, k. Thus in general $E = Q/(A * k * \varepsilon_0)$. For instance, if there is (pure) water between the plates (at standard room temperature and pressure), k would be about 79 (a lot bigger than 1!).

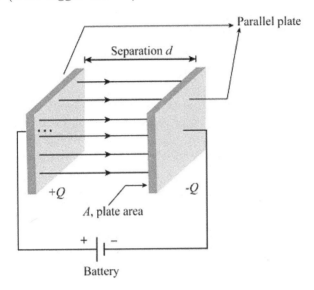

Combining our two equations for E, we find that $V = d * E = d * Q/(A * k * \varepsilon_0)$. Then, we can finally find the capacitance, $C = Q/V = A * k * \varepsilon_0/d$. We find that the capacitance *does* depend on the details of the object. For our light bulb, maybe the combined area of the wires in the bulb is 1 square inch, and the distance between the wire coils is 0.1 inch, and there's vacuum in the bulb. Plugging in the numbers, this gives a capacitance of about 2.25×10^{-12} Coulomb/Volt, or Farads (F). (We get tired of saying C/V so we give capacitance the unit name Farads.)

We found in the last chapter that our wire has 1.5 C of free charge in just about 0.1 mm. And we only need a charge excess of 1.5 V $\times$ 2.25 pF, so the charge in our wire only needs to travel about 0.1 mm $\times$ $2.25 \times 10^{-12} = 2.25 \times 10^{-16}$ m! That's about a million times smaller than the diameter of an atom! The charges in the wire might move slowly (about 0.1 mm/s), but they don't need to move far, usually.

We might estimate that it would only take about 2.25 ps (trillionths of a second) for our needed charge to move into place.

The RC Circuit

Let's calculate more precisely how the voltage changes vs time for a circuit with capacitance. Have a look at the "RC circuit" figure below. When the switch is closed, the current must flow through the resistance R, which can then "charge" the capacitance C, until the voltage across C, V_c, equals the voltage across the battery, V_s. If we go back to the definition of capacitance, it says that $C = Q/V$, or $Q = CV$. Taking the derivative of both sides vs time gives $dQ/dt = C\,dV/dt$. But dQ/dt is also just what we have called current, I! The change per second of the excess charge on the capacitor is the same thing as the current going to the capacitor. So $I = C dV/dt$. Note that the current is the same everywhere in the circuit since everything is in series.

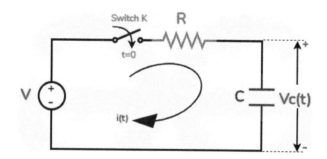

By Ohm's law, the current also equals V/R, where V is the voltage difference across R. So $I = (V_s - V_c)/R$. Combining the two equations for I gives $(V_s - V_c)/R = C\,dV_c/dt$. Integrating, we find that $V_c = V_s * (1 - \exp(-t/(R*C)))$. The factor $\tau = R*C$ (often called the time constant, which I find a paradoxical name) is an amount of time! So if you have a 1 k Ohm resistor and a 1 uF capacitor, $RC = 1$ ms. Does it make sense that $R*C$ is time? Well $R = V/I$ and $C = Q/V$. $RC = V*Q/(I*V) = Q/I$. But I is Q/s, charge per second. And $Q/(Q/s) = s$! Drawing V_c vs t, we see a curve like this:

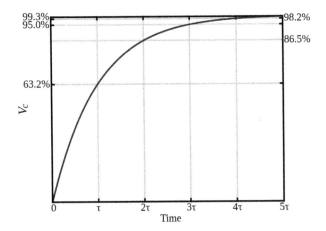

The voltage across the capacitor starts at 0 of course, but then rises "exponentially" towards 100% of Vs. After one $\tau = RC$ of time, the voltage V_c has risen to $1-1/e$, or about 63.2%. And each $\tau = RC$ afterwards V_c will rise approximately another 63.2% of the distance left to V_s. After about 5τ, the capacitor is "fully charged". Discharging the capacitor, by setting $V_s = 0$ (i.e. replacing the battery with a wire), would also just fall exponentially like $V_c = V_s * \exp(-t/\tau)$. The larger the resistance, the longer it takes to charge or discharge the capacitor. This makes sense, since when R is larger, the current into or out of the capacitor is smaller. And it takes current to deliver the needed charge to or from the capacitor. So the real answer for our flashlight is that the voltage would rise to ~63.2% of V_s after $\tau = RC = 1\Omega * 2.25\,\text{pF} = 2.25\,\text{ps}$, pretty fast!

Displacement Current

This charging (or discharging) capacitor is a bit of a strange circuit. We have current flowing from the battery through the wire, into the capacitor, no current inside the capacitor, then suddenly (the same) current magically reappearing on the other side of the capacitor, and flowing back to the battery. It "feels" like there should be some sort of current that is conserved and thus also flowing through the capacitor. Also, the fact that there's no current in the capacitor causes a big problem for Ampere's law, which says that the integral of the magnetic field, B, around a closed loop equals μ_0 times the integral of the current flowing through any surface with that same closed loop as its boundary. The integral of the current flowing through the surface should be the same no matter what surface we choose, as long as the boundary stays the same. But for the capacitor circuit, we could take the surface to cut through the wire, as drawn below, but we could also bend it (keeping the boundary the same) to make it cut only through the middle of the capacitor. In the first case there would be current going through the surface, but in the second case there would be none!

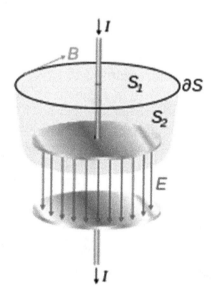

In order to restore consistency, Maxwell proposed the "displacement current", which must be equal to the current charging the capacitor, $\tau_0 \, dE/dt$. Then the "total current" at any point in space is always the normal charge current plus the displacement current. The integral of the "total current" through any surface with the same boundary will then always be the same. While it may seem at first like a small fix, this addition makes it possible for changing electric fields to generate magnetic fields, in empty space, with no actual charges. When combined with Faraday's law, that a changing magnetic field generates an electric field, Maxwell realized that electromagnetic waves could self-sustain themselves and propagate through vacuum — and he could even calculate that those waves would travel at a velocity equal to the known velocity of light!

Reactance

To further understand time-dependent circuits, it's very useful to study just a single frequency at a time. In other words, let's input a voltage which varies like $A * \cos(wt)$, instead of the constant voltage from the battery. A is the amplitude of the voltage and w is the angular frequency (f), so $w = 2\pi f$. Below we draw two circuits, each driven by a voltage with an amplitude $A = 1\,\text{V}$ and frequency $f = 100\,\text{Hz}$, so $V = \cos(2\pi * 100 * t)$. For the resistor, nothing very special happens — the current through it is just $I = V/R = A * \cos(wt)/R$. Such a current, which alternates the direction it goes each half cycle, is called *alternating current* (AC), as opposed to the direct current (DC) we had studied before, which was driven by a constant voltage.

For the capacitor, it is a little more interesting. The current is $I = C * dV/dt = -C * w * A * sin(wt)$, because the derivative of $\cos(wt)$ with respect to t is $-w * \sin(wt)$. We could also write the current as $I = C * w * A * \cos(wt + \pi/2)$, since $-\sin(x) = \cos(x + \pi/2)$. So what we've found is that the (displacement!) current through a capacitor is equal to $C * w$ times the voltage across the capacitor, but $\pi/2(90°)$ ahead in phase.

Since there is a voltage across the capacitor and a current through the capacitor, it sort of makes sense to define the "resistance" of the capacitor, "R" $= V/I$. It is $A * \cos(wt)/(C * w * A * \cos(wt + \pi/2))$, which is *nearly* equal to $1/(wC)$, aside from the fact that the current is 90 degrees ahead in phase. We call this "reactance" since it's not really resistance, and we have to always remember later on whenever we use it that the current is 90 degrees ahead of the voltage in phase. This turns out to be a very useful concept — to think of capacitors as resistors whose resistance depends on frequency. As the frequency goes to zero, the reactance goes to infinity, so capacitors completely block DC current, which makes sense. But as frequency goes to infinity, the reactance goes to zero, so capacitors are like wires for high frequencies! How large a frequency would we need such that the $1\,\mu F$ capacitor in our example below has equal reactance to the resistance of the $1\,k$ resistor? Well $1/(w * 1\,\mu F) = 1\,k\Omega$, so $w = 1/(1\,\mu F * 1\,k\Omega) = 1000 = 2\pi f$, so $f = 1000/(2\pi) \approx 160\,Hz$, not that big!

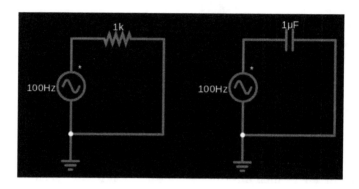

Complex Impedance

Since the capacitor has a sort of resistance (reactance), it means we should be able to add them together, just like two resistors in series or parallel. Since the sum will be some combination of resistance and reactance, we need a new word for that — *impedance*, which we also abbreviate as Z (since I was already taken!). Note that Ohm's law still holds by definition, so $V = I * Z$. If the resistor and capacitor are in series, for example, the total impedance should be something like $R + 1/(wC)$. But this isn't quite right, because the current through the capacitor needs to be 90 degrees ahead of the voltage across it. We need a way to "add" two objects that accounts for the phase of the object as well as its magnitude. Fortunately, we know two ways to do this.

One way is to treat the objects as 2D vectors, with a magnitude and phase angle, and add them vectorially. These are called *phasors* in electronics, and they're ok in some circumstances. But they end up being cumbersome — you end up taking tangents of angles all the time, etc. And the product of two 2D vectors is either a dot product, which gives a scalar, or a cross-product, which gives a vector perpendicular to the two original vectors (but what is that 3rd dimension in this case?). Neither way of defining the product gives back another vector in the original 2D space.

The other way is to use complex numbers, which have no trouble with multiplication (or division) and no annoying angles to deal with! We can think of the input voltage as the *real* part of $A*e^{iwt}$, which is the real part of $A*\cos(wt)+i*A*\sin(wt)$, which is just $A*\cos(wt)$. The current through the capacitor will be (following our logic above, but this time using complex notation) the real part of $I = C * dV/dt = C * i * w * A * e^{iwt}$. So $Z = V/I$ is just $1/(i*w*C)$! The fact that the current is 90 degrees ahead of the voltage in phase is all captured in the simple factor of $1/i$. This "rotates" the complex number by 90 degrees in the complex plane. Using complex notation, the total impedance of a resistor and a capacitor in series is just $Z = R+1/(i*w*C)$.

Voltage Divider with a Capacitor

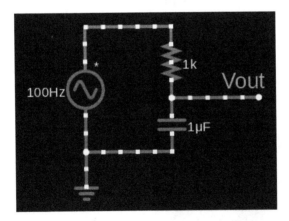

We can use this impedance idea to easily calculate circuits involving capacitors and resistors (and inductors, as we'll see in Chapter 3). For instance, what is the output voltage of the circuit below, which is the voltage across the capacitor? If you recall the voltage divider from Chapter 1, it is identical to this circuit except that here we have replaced the bottom resistor (R_2) with a capacitor (C). The output voltage from the voltage divider was $V_{out} = V_{in} * R_2/(R_1 + R_2)$. To find the output voltage from this circuit, we just replace the impedance of R_2 (which is R_2) with that of the capacitor, $1/(i * w * C)$. So $V_{out} = V_{in} * 1/(i * w * C)/(R_1 + 1/(i * w * C)) = V_{in}/(1 + i * w * R * C)$. Let's analyze this expression. First, note that $R * C$ is the "time constant" we saw above for charging the capacitor, and it has units of seconds. When $w \gg R * C$, at high frequencies, V_{out} goes to $V_{in}/$infinity, which is zero. This makes sense because capacitors are like wires at high frequency, and a wire connected to ground will be at 0 V. When $w \ll R*C$, at low frequencies, V_{out} goes to $V_{in}/1$, which is V_{in}. This also makes sense, because at low frequencies capacitors have infinite resistance, and so the output is only connected to V_{in}, with a resistor in between.

When $w = 1/(R * C)$, we get $V_{out} = V_{in}/(1 + i)$. What does that mean?! Well, $1 + i$ has a magnitude of $\sqrt{2}$, so the magnitude of V_{out} is $V_{in}/\sqrt{2}$. You may sometimes hear this referred to as -3 decibels (dB) in magnitude, and this frequency called $f_{-3\,dB}$. A dB is defined as $20 * \log_{10}(V_{out}/V_{in})$. So $V_{out}/V_{in} = 1/\sqrt{2}$ gives $20 * \log_{10}(1/\sqrt{2}) \cong -3\,dB$. A factor of 2 reduction in amplitude would be about -6 dB. And a factor of 10 reduction would be (exactly) $-20\,dB$. Those are the dB values worth remembering. Corresponding increases in amplitude would have positive dB values. Also keep in mind that because the dB scale is a log scale, you *add* the dB values of two successive filters in order to find their combined effect. For example, if you had two filters that each reduced the amplitude by a factor of $\sqrt{2}$, and thus each had a dB value of $-3\,dB$, to find the total dB value of both filters combined in series is $-3\,dB + -3\,dB = -6\,dB$.

We can also calculate the phase of V_{out} relative to V_{in}. The phase, or angle, of $1 + i$ is 45 degrees, so V_{out} is 45 degrees behind V_{in}. It makes sense that it would be behind V_{in}, because in the limit that R is small, we just have a capacitor, and we knew from above that in that case the current is ahead by 90 degrees. (It gets a bit confusing which way the phase goes, ahead or behind, but you can almost always figure it out by just thinking about it!)

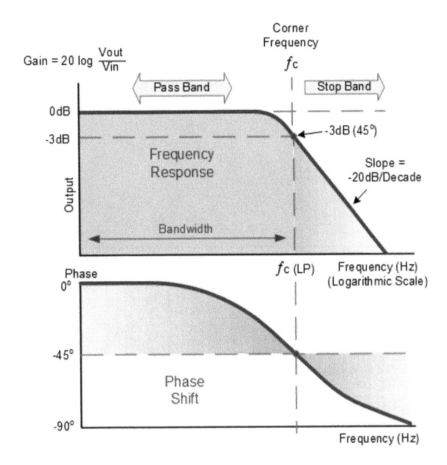

What we have created with this simple resistor and capacitor is a *low-pass filter*. This means that low frequencies are passed through the filter from the input (V_{in}) to the output (V_{out}). But high frequencies (roughly those above $f_{-3\,\text{dB}}$ are attenuated. This is an incredibly useful circuit, and we'll use it countless times going forward. Of course, we can just as easily make a *high-pass filter*, with opposite behavior vs. frequency. Just switch the resistor and the capacitor, so the resistor is on the bottom. It would have a response like this figure below. Note that the phase for the output voltage of a high-pass filter is ahead of that of the input voltage.

$$\text{Gain (dB)} = 20 \log \frac{\text{Vout}}{\text{Vin}}$$

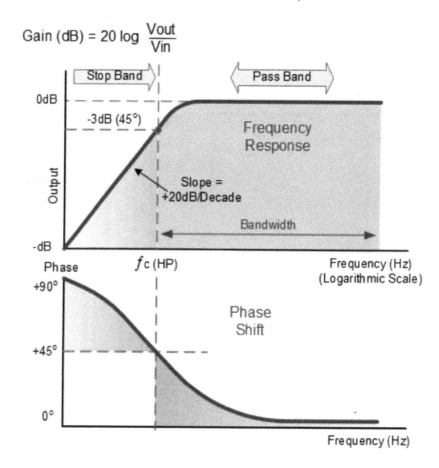

Band-pass Filter

How would you make a *band pass* filter that reduces the amplitude of low frequencies, say below 160 Hz, but also reduces that of high frequencies, say above 1600 Hz. The basic idea would be to put a low pass filter on the output of the high pass filter, as shown below. (You could instead make a *notch* filter by putting a high pass filter in parallel with output of a low pass filter.) But unless you're careful, this won't work as expected! For instance, for V_{in} of 160 Hz and 1.5 V, V_{out} should be about $V_{\text{in}}/\sqrt{2} = 1.063\,\text{V}$, since the $f_{-3\,\text{dB}}$ of the high pass filter is about 160 Hz, and the $f_{-3\,\mathrm{dB}}$ of the following low pass filter is at 1600 Hz which is too far above 160 Hz to have any effect. But a quick simulation of the below circuit in fact gives V_{out} of just 0.66 V!

The problem is we haven't been careful about output and input *impedance*, which are the complex generalizations of output and input resistance. The output impedance of the high pass filter is that of the resistor and capacitor in parallel, just like for the voltage divider from the previous chapter. So the output impedance is $1/Z = 1/R + iwC$, so $Z = 1/(1/R + iwC)$, where Z is impedance as you'll recall. Z is frequency dependent, because the impedance of the capacitor is frequency dependent. But in the "worst" case, when it is highest, when w goes to zero, it is just R. Going to higher frequencies will just lower the output impedance. The input impedance of the following low pass filter is that of the resistor and capacitor in series (again just like for the voltage divider), so $Z = R + 1/(iwC)$. Again the "worst" case, when it is lowest, when w goes to infinity, it is just R. Going to lower frequencies will just raise the input impedance.

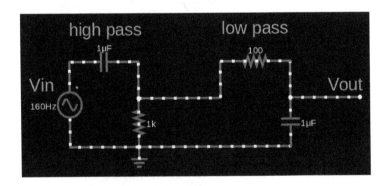

To be safe, we want to make sure that the input impedance of the low pass filter is at least 10× that of the output impedance of the preceding high pass filter, for all frequencies. The only way to do this is to require that the resistor of the low pass filter is 10× the resistance of that in the high pass filter, as shown below. Indeed, if we do this the output voltage is 1.058 V, quite close to the 1.063 V expected.

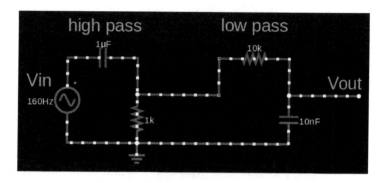

AC Power

What about the power used by alternating voltages and currents? We had said before, for DC voltages and currents, that $P = V * I$. But here the voltage and current are changing all the time, so which voltage and current do we use? The power at each instant of time is given by the current and voltage at that instant in time. What we need to do to find the average power being used is to integrate the power over a complete cycle. Recall that the voltage is $V = A * \cos(wt)$ and the current for the circuit with just a resistor is $I = V/R = A * \cos(wt)/R$. So, $P = V * I = A^2\cos^2(wt)/R$. The integral of $\cos^2(wt)$ over a cycle is $\frac{1}{2}$. (You can easily prove this by remembering that $\sin^2(x) + \cos^2(x) = 1$, and the integral of $\sin^2(x)$ and $\cos^2(x)$ must be the same, since they are just shifted by a phase from each other.) So, the power is $P = A^2/(2 * R)$.

Remember that A is the *peak* amplitude of the voltage, the largest (and smallest) that the voltage gets. Another way to write the calculation of power is $P = (A/\sqrt{2})^2/R$. So, if we had taken $A/\sqrt{2}$ for voltage and $(A/\sqrt{2})/R$ for current, we actually could have used $P = V * I$ to calculate the power. $A/\sqrt{2}$ you may also recognize as the *root-mean-square* (RMS) voltage. It turns out this works in general, for any voltage, constant or non-constant, sinusoidal or of any shape at all — as long as we are just talking about resistive loads. (For DC, the RMS is equal to the peak voltage, so we were safe using peak voltage for that special case before.) The type of voltage given by the common household outlet (in the US) is 120 V AC RMS. The peak voltage is $\sqrt{2}$ times larger, about 170 V.

When capacitors (or inductors, discussed in the next chapter) are involved, power equals the *real part* of $V * I = V * I * \cos(\phi)$, where ϕ is the phase difference between the voltage and the current. Note that for V and I 90 degrees out of phase, for a purely capacitive load, $\cos(\pi/2) = 0$, so $P = 0$! Capacitors (and inductors) use no power, only resistors do! (Or at least perfect ones don't ... real capacitors and inductors have some small resistance, often called effective series resistance (ESR)). Resistors have "friction" and produce heat. Capacitors (and inductors) have no friction — the energy that goes into them when their voltage (or current) is rising is later returned in full when their voltage (or current) is falling. The energy is stored, by the way, in the electric (or magnetic) field.

Lab: Capacitance and Filters Lab

A few things to keep in mind about capacitors

1. Big caps (>1 uF) are often electrolytic (or similar), and this type of cap is "polarized", meaning it must have the $+$ side at higher V than the $-$ side or "bad things" will happen (like the cap can burst!).

2. Like all components, capacitors have voltage limits. For caps it is often as low as ~ 25 V, but you can get caps that can withstand ~ 1 kV or more, while some may only support ~ 3 V! Read the label on the part, or consult the datasheet, for every component before using it! Also beware that capacitance can decrease significantly when at higher voltage, especially for multilayer ceramic caps. A 10 uF cap might only have 1 uF of capacitance near its rated voltage!

3. You can measure capacitance using a multimeter (or at least many multimeters support this feature). How do you think it measures capacitance?

4. There are lots of different kinds of capacitors which have various properties in addition to their capacitance and voltage tolerance: https://en.wikipedia.org/wiki/Capacitor_types. Some are more precise and stable with temperature or time, have less internal resistance or inductance, etc. These differences won't matter

for this exercise, but may be critical to your application in the future!

Oscilloscope

To accurately measure voltages that change in time, we need to use an *oscilloscope*, not a multimeter. The oscilloscope is typically used to plot voltage vs. time. Here is an excellent introduction to using an oscilloscope: https://learn.sparkfun.com/tutorials/how-to-use-an-oscilloscope.

Just a few things to keep in mind:

1. Use oscilloscope probes for inputting signals to the oscilloscope. They are not just wires! (If you don't believe me, measure the resistance, using your multimeter, of the scope probe from the probe tip to the BNC center pin on the other end.) Using a BNC cable to directly connect to the scope input from a circuit output or another piece of equipment like a signal generator (discussed next) will work well only at low frequency or when you have *matched impedance* between the output, the cable (usually 50 Ω), and the scope input (usually 1 MΩ). For now, simplify your life and just use a probe as a scope input. We will worry about matching impedance and high frequency measurements later.

2. Don't use a scope probe for anything other than as an input to the oscilloscope. For instance, don't connect it to the output of a signal generator and try to use it to connect from there to your breadboard! Again, scope probes are not just wires!

3. *Compensate* your probes, using the built in square wave output of the oscilloscope, before measuring with them. This is usually stable for days or weeks or even years once calibrated, but it's good practice to check the compensation now and then, especially if the probe has been moved around.

4. Probes generally have two *modes*, 1× and 10× mode. In 1× mode, the voltage at the probe tip (relative to ground of course) is equal to the voltage input to the scope. But in 10× mode, the voltage at the scope input is 10× LESS than the voltage at the tip. Scopes allow you to correct for this factor by telling the scope that your

probe is in 10× mode, usually through an option in the channel menu. The advantage of 10× mode (in addition to allowing you to measure larger voltages) is that the *bandwidth* is also ∼10× larger. For any measurements above ∼1 MHz, make sure to use 10× mode.

5. Oscilloscope channels can be either in DC or AC mode, set through an option in the channel menu. When in AC mode, a high-pass filter is used to remove all frequencies below a few Hz, including the DC (0 Hz) voltage. That can be really confusing — you are changing some voltage, measuring it with the scope probe, and always seeing 0 V! You were in AC mode. The advantage of AC mode is that you can measure just the *fluctuations* of a voltage, such as how much noise there is at a particular point in the circuit, without being swamped by the large constant voltage present.

6. Ground the scope probe using the alligator clip on the *ground lead* (or a *ground spring* for high frequency measurements, roughly >1 MHz) to the ground of your circuit. This is the reference ground with respect to which you are measuring voltage at the probe tip. If you don't attach the ground clip, the probe will be measuring voltage with respect to the ground of the oscilloscope, almost always defined as the ground pin of the 3-prong plug with which the oscilloscope is plugged into the wall for AC power, which is then connected to *Earth ground* via the electrical system of the building. Depending on how your circuit is powered or connected to the outside world, it may also have its ground connected to Earth ground, or maybe not. If not, we'd call that an *isolated* circuit. Maybe it's powered by a battery for instance. Then your scope probe is measuring voltage in your circuit with respect to the unrelated ground reference of the oscilloscope. You will measure nonsense. Or, maybe they do share the same ground reference. For instance, if you are using a signal generator to feed input to your circuit, the ground of the signal generator is likely connected to Earth ground. In this case, at least your scope probe with unattached ground lead would be measuring with respect to the same reference voltage, but imagine the terrible circuit you've created — current must now flow from your signal generator output, through your circuit, into your probe tip, through your

scope, into the wall, through the Earth (or maybe just through the building's wiring if they're on the same electrical feed, but still), through the wall, and into the ground of the signal generator. Such a non-ideal path will have considerable resistance, large inductance, and be a great antenna for receiving low frequency electromagnetic noise, including the 60 (or 50 in many countries) Hz AC power frequency. Attaching the ground lead of the scope probe to your circuit allows the current to flow simply, from the ground of your scope through the ground of your circuit to the ground of the signal generator.

7. Do NOT attach the ground of the probe to a voltage other than ground! Potentially large current will flow from the voltage source through the low resistance "ground" path of the scope. You could destroy the oscilloscope! https://www.youtube. com/watch?v=xaELqAo4kkQ. The oscilloscope ground is NOT a *floating ground*, like the ground of the (battery powered) multimeter! On the multimeter you can put the black probe on whatever voltage you want. That is not true for most oscilloscopes. To do so, you need a special *isolated* scope or probe. Or, if the circuit itself is isolated from Earth ground (e.g. a battery powered circuit), then you're safe and can put your scope ground anywhere in that circuit you like. You will then be forcing that point in your circuit to be equal to Earth ground, but if your circuit is otherwise isolated, that's OK.

Signal generator

To create a time varying voltage, we need a *signal generator* (sometimes also called a *function generator*). These usually have two channels, each of which outputs (with low output impedance, usually $50\,\Omega$) a voltage which can vary in time. You can set the type of *wave* output, like a sine wave, square wave, short pulses, etc. On fancier ones you can even define your own wave, like have it play back a particular pulse sequence vs time. You can also vary the voltage amplitude and voltage offset and frequency of those waves. The two outputs can furthermore be synced, such that the *phase* of the second output with respect to the first output can be set to a given value.

Exercises

1. Use a function generator and an oscilloscope to measure several waveforms applied to a single resistor on the breadboard, showing voltage vs time. Measure the amplitude and frequency of each waveform using the oscilloscope and compare to that claimed by the signal generator. Trigger on the rising edge of each waveform so that the signal is stable.

2. Make a low-pass RC filter and study its frequency response. In other words, measure the RMS output voltage of the filter as a function of the input signal frequency. Does the output frequency equal the input frequency?
   Find the measured $-3\,\mathrm{dB}$ cutoff frequency and compare to a calculation of the $-3\,\mathrm{dB}$ frequency, given the component values used. Compare the shape of the frequency response to that in simulation. Try all this for a few different component values and thus a few different cutoff frequencies.

3. Add another capacitor in parallel, and then in series, to the original capacitor. Measure total capacitance and compare it to theory.

4. What is the phase shift of the filter, for frequencies below, at, and above $f_{-3\,\mathrm{dB}}$? Make a plot of phase shift vs frequency and compare it to simulations.

5. Repeat the above steps for a high-pass RC filter.

6. Combine the low and high pass filters to make a band-pass filter, and compare its frequency response to simulations. (Beware the 10% rule for output to input impedance between circuit stages!) Also compare the phase shift of this band-pass filter vs frequency to simulations.

Chapter 3

Inductance and Power

Let's go back to our flashlight and think about yet one more thing that happens when we turn it on. For now we'll assume the light bulb (and wires) have no capacitance. That means the voltage can rise immediately from 0 V to 1.5 V on one side of the bulb relative to the other. Will the current immediately go from 0 A to $I = V/R = 1.5 \text{ V}/1 \text{ } \Omega = 1.5$ A?

Inductance

There's an important effect we've been ignoring up to now. As the current increases (from 0 A to 1.5 A eventually), the magnetic field caused by that current will also increase. For instance, as shown in the following picture, the magnetic field will form in circles around the wire, as given by Ampere's law and the right-hand rule. As the current increases, the magnetic field increases, which means we have a *changing* magnetic field through the loop defining our circuit. But this has an effect on our circuit! Faraday's law (one of Maxwell's equations) says that the voltage generated around a loop is given by $V(t) = -dB/dt$, the (negative of) the rate of change of the magnetic field (flux) through the loop. The magnetic field is proportional to the current, so $-dB/dt = LdI/dt$, where the constant of proportionality is L, the *self inductance* of the circuit. The larger the inductance, the larger the magnetic field, for a given current. Putting these together, the final result is that $V(t) = LdI/dt$, a voltage proportional to L is induced *opposing* the increase of current in a circuit. Circuit elements that intentionally have large L are called *inductors*, but all circuit elements will have some small inductance, like it or not, just as all circuit elements also have some capacitance.

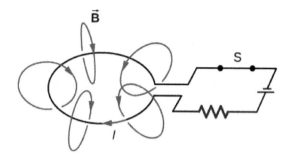

The RL Circuit

So, we actually need to consider two sources of voltage when we turn on the flashlight (still ignoring capacitance for the moment), the battery voltage (V_B), and also $L\ dI/dt$. Thus by Ohm's law, we have $V_B - L * dI/dt = I * R$. We know how to solve this equation to get $I(t)$ since it's the same as for the $V(t)$ for a capacitor. $I(t) = (V_B/R) * (1 - e^{-t/\tau})$, where $\tau = L/R$ is the time constant for the circuit. Remember that L/R is a time, just like $R * C$ for a capacitor. The current rises towards the final value, V_B/R (1.5 A for $V_B = 1.5$ V and $R = 1\ \Omega$), but only gets to $1 - e^{-1} \sim 63.2\%$ of V_B/R after τ seconds.

The units of inductance, L, are V/(A/s), which we definitely got tired of saying, so we renamed the Henry (H). So, 1 H is the inductance of an inductor that generates 1 V (opposite to) the current changing at a rate of 1 A/s. To give an idea of some practical numbers, a straight wire of length 10 cm and 1 mm wide (like the ground wire on your oscilloscope probe!), has $L \sim 100$ nH. You can experiment with different geometries with an inductance calculator. If our flashlight had this inductance, 100 nH, from the combined effects of the wires and bulb, we would have $\tau = L/R = 100$ nH/1 Ω = 100 ns. This is still pretty fast, which is why you've never worried about waiting for your flashlight to turn on, but it is much slower than the 2.25 ps, required to charge the capacitance of the light bulb!

Inductance grows with size, since the total magnetic field grows proportionally to the length of the wire carrying the current. But capacitance decreases with size, since the electric field from one side of the capacitor on the other is smaller if they're farther apart. To make voltages and currents change quickly, we'd like small inductance and small capacitance, but it's impossible to make them both very small at the same time. If we make the flashlight smaller, to minimize inductance, we end up raising capacitance. And if we make the flashlight bulb larger, to minimize capacitance, we end up raising inductance. The proper geometry that minimizes both, for a given circuit, is complicated, and something that electrical engineers designing integrated circuit chips (and other devices) work hard on. But we can already see that decreasing the size of the flashlight by a factor of ~100 would already be a nice first step. The inductance would then be roughly 1 nH, so $\tau = 1$ ns, and the capacitance τ might rise to 225 ps, so both the current and the voltage can rise quickly, and our flashlight might be able to turn on in just ~1 ns. This is roughly the fastest we can easily make voltages and currents turn on and off, which is why electronic computer processors and other fast devices run at ~GHz speeds.

In other cases, we will want to have *large* inductance. This is usually done by coiling a wire around, so that it makes N turns. The magnetic field from each turn adds linearly, and that flux passes through N times as many loops, so the inductance grows quadratically, giving an inductance N^2 times larger than for a single loop. You can typically make inductances of ~10 uH or so this way with ~100–1000 turns. For larger inductances, the trick is usually to use an iron core inside the coil. The relative permeability of Iron is ~5000, easily giving inductances around ~50 mH. Don't forget that real inductors also can have significant resistance, from all those loops of wire, and also capacitance.

Complex Impedance of an Inductor

We'll now consider applying a single frequency AC current to an inductor, $I(t) = A * \cos(wt)$, or the real part of $A * e^{iwt}$. Using the equation above, $V(t) = L \, dI/dt$, we find that $V(t) = Liw * I(t)$. So, the impedance of an inductor, V/I, is iwL. We find that the impedance of an inductor is proportional to L and *grows* with frequency, whereas for the capacitor the impedance *decreases* with frequency.

Voltage Divider with an Inductor

Can we use an inductor (instead of a capacitor) to make a low-pass filter? Sure! Since the impedance of the inductor goes to zero as the frequency goes to zero, we want it on the top of the "voltage divider", and a fixed resistor on the bottom, as shown in the following figure.

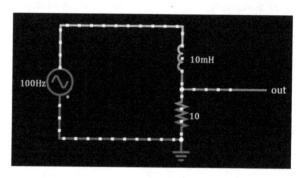

Again we'll use our voltage divider equation, $V_{out} = V_{in} * R_2/(R_1 + R_2)$, but R_1 is now an inductor with impedance iwL, and $R_2 = 10\ \Omega$ which we'll just leave as R. So $V_{out} = V_{in} * R/(iwL+R) = V_{in}/(1 + iwL/R)$. You should recognize this as having identical form to what we found for the capacitor low-pass filter in Chapter 2, $V_{in}/(1 + i * w * R * C)$, except here the time constant $\tau = R * C$ has been replaced with the time constant for an inductor and resistor, $\tau = L/R$. The output impedance is that of the inductor and resistor in parallel, so $Z = 1/(1/R + 1/iwL)$, which goes to R as w goes to infinity, and goes to zero as w goes to zero. So, just as for the RC low pass filter, the output impedance is at most R. The input impedance is $Z = R + iwL$, which goes to R as w goes to zero, and goes to infinity as w goes to infinity. So again, just as for the RC low pass filter, the input impedance is never smaller than R.

You may ask, why ever bother using an LR low-pass filter instead of an RC low-pass filter, if they do the same thing? The choice of which to use comes down to practicality and engineering.

For example, to get a low cutoff frequency you would need a large $\tau = L/R$ or $\tau = RC$. Given that it's expensive to make a large capacitor or inductor to make C or L large, it's easier to just reduce R, and use the $\tau = L/R$ of an inductor. There are also considerations from input and output impedance, which have opposite frequency relations for RC and LR filters.

Improved Band-Pass Filter

The inductor does allow us to make an improved band pass filter. Adding a capacitor, in series, to our LR low pass filter above will block low frequencies, where the impedance of the capacitor is large. This so-called RLC series band pass filter is shown below.

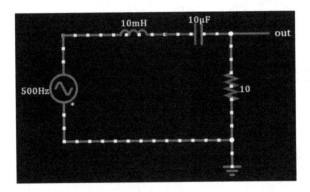

To analyze the effect of this filter, we again want to know the *gain*, that is $V_{\text{out}}/V_{\text{in}}$. We have $V_{\text{in}} = I * Z = I * (iwL + 1/iwC + R)$ and $V_{\text{out}} = I * R$. So $V_{\text{out}}/V_{\text{in}} = R/(iwL + 1/iwC + R)$. This is not particularly illuminating, so a graph of the *magnitude*[1] of $V_{\text{out}}/V_{\text{in}}$, $\frac{R}{\sqrt{\left(Lw - \frac{1}{Cw}\right)^2 + R^2}}$, is also shown in the following figure, for $R = L = C = 1$, as a function of w.

---

[1] Remember that to calculate the magnitude of a complex number, z, you take the sqrt of the number times its complex conjugate, z^*. So, $|z| = \sqrt{z\,z^*}$. And remember that to find the complex conjugate, you just replace i with $-i$ everywhere in the number or expression.

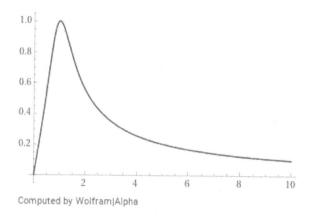

Computed by Wolfram|Alpha

Note that the gain goes to $R/R = 1$ at the frequency when $iwL = -1/iwC$, so the impedance of the inductor is exactly equal to, but opposite in phase to, the impedance of the capacitor, and the two terms cancel. This *center* frequency is $w_0 = \sqrt{1/(LC)}$. We also care about the width of the peak, quantified as the distance from where it is $1/\sqrt{2}$ on the left, to where it is $1/\sqrt{2}$ on the right. These two points are at:

$$\omega_C = \pm\frac{R}{2L} + \sqrt{\left(\frac{R}{2L}\right)^2 + \left(\frac{1}{LC}\right)}$$

so the width from one point to the other, often called the *bandwidth* (*bw*), is R/L. A final important measure of the shape is how peaked it is, often called the Q *value*, given by the ratio of the center frequency to the bandwidth. For this filter it is $Q = w_0/bw = L * w_0/R = L * \sqrt{1/(LC)}/R$, which simplifies to $\sqrt{L/C}/R$. In the graph we plotted, for $R = L = C = 1$, $w_0 = 1$ and $bw = 1$ and thus $Q = 1$. To get a narrower filter that selects a smaller range of frequencies, we need larger Q. We can achieve this by using a larger inductance (and a correspondingly smaller C, to keep w_0 constant), or by reducing R.

There are some practical limitations to how large we can make Q though. Remember that the inductor itself also has a resistance, called the *effective series resistance* (ESR), because it is made of wire, which has some resistance. This more realistic circuit is shown below. A larger inductance is hard to create without also increasing the ESR proportionally. That resistance will add to the usual R, and spoil the Q of the filter. Also beware that once inductor ESR is included, the filter will no longer have a gain $= 1$ at w_0. Remember that at

w_0, the impedance of the inductor cancels that of the capacitor. So at w_0 we are just left with a voltage divider, with ESR on top and R on the bottom. So the gain at w_0 will realistically be $R/(R + \text{ESR})$, which would be $10 \ \Omega/(10 \ \Omega + 2 \ \Omega) = {\sim}0.83$ for our example shown.

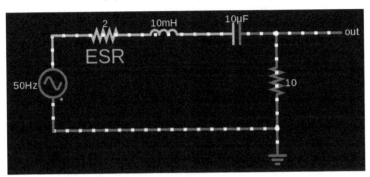

Transformers

An important use of inductance is in making a *transformer*, a device used for changing the RMS voltage of an AC voltage. It has an inductor on the input side with AC voltage V_{in} and current I_{in}, and another inductor on the output side with AC voltage V_{out} and current I_{out}. The trick is that the magnetic flux from the first inductor is made to also flow through the second inductor, usually by placing a high magnetic permeability material, like Iron, through them. (We could say that mutual inductance is maximized, but this is a concept we'll discuss later.) The voltage around each turn of wire is proportional to dB/dt, the rate of change of flux through the turn, and this dB/dt is the same on both the input and output sides. That would mean that V_{out} equals V_{in}, except that there can be a different number of turns of wire on the output vs the input side. So the total V_{out} is $(N_{\text{out}}/N_{\text{in}}) * V_{\text{in}}$, N_{out} and N_{in} are the number of turns of wire around the flux on the output and input sides, respectively. For example, if there are 10 turns on the input side and 100 turns on the output side, and we feed in 120 V (RMS!) on the input side, we will get $V_{\text{out}} = (100/10) * 120 \ \text{V} = 1200 \ \text{V}$ (RMS) on the output side. Of course, energy must be conserved, and moreover the power coming in must equal the power going out, at each instant in time. So $P = V_{\text{in}} * I_{\text{in}} = V_{\text{out}} * I_{\text{out}}$. Thus if V_{out} increases by a factor of 10, I_{out} must correspondingly *decrease* by a factor of 10. Also note that the output side is *isolated* from the input side, meaning it is not directly connected electrically. In particular, the ground reference point on the output side can be very different from the ground reference point on the input side — they do not share a ground connection. This can be a very useful feature, as we'll see.

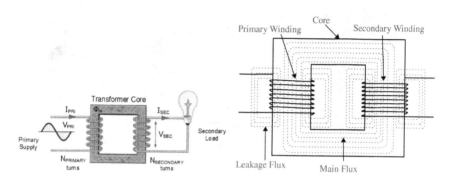

Transformers essentially make power distribution over long distances possible. The AC voltage from a power plant is "stepped up" to high voltage, where it is then distributed through wires for hundreds of miles with low losses, because power lost in wires is $P = I^2R$, and I is small, because V is big, and $P = VI$. Then the voltage is "stepped down" near homes to safe voltages. The result is 120 V AC RMS coming out of the wall.

Voltage Rectification

For many of our circuits, however, we will need DC power, not AC like that coming from the wall. Even for charging our phones and laptops (and cars!), we need DC voltage and current. Batteries charge by applying a constant DC voltage to them — an AC voltage would be pushing the ions in the battery back and forth, damaging the battery. So how can we create, say, 5 V DC from a 120 V AC one?

First, it's common to use a transformer to go from 120 V AC to 5 V AC. But then we need to turn this 5 V AC (RMS) into 5 V DC. A good start would be to remove the negative part of the voltage. For this we'll make use of a new component, the *diode* — a semiconductor component that only allows current to flow in one direction. More precisely, it only allows current to flow with small resistance once the voltage across it is above the "diode drop" voltage, usually about $\sim$0.6 V in silicon, but determined by the semiconductor material band-gap. Diodes don't allow current to flow "backwards" until the "breakdown voltage" of $\sim$100 V, also determined by the semiconductor material. There are many types of diodes with many special properties, such as smaller or larger voltage drops, precise voltage drops, precise breakdown voltages, etc.

There is even the ubiquitous *light-emitting diode* (LED), which emits light of a particular color (frequency). The frequency of the light emitted is proportional to the LED voltage drop. Each electron that passes through the LED can release a "photon" of light, and the frequency of that photon is given by $E = h * f$, where h is *Planck's constant* and f is the light's frequency. For instance, if the voltage drop is 2 V, the frequency of light is $f = E/h = 2$ V$/4.14 \times 10^{-15}$ eV s) $= 480$ THz, which is red light. (An *electron volt* (eV) is the energy of a single electric charge with a potential energy of 1 V.)

We can see the effect of a diode in the circuit below after the transformer. The 5 V RMS AC voltage made by the transformer has a max and min voltage of $5 * \sqrt{2} = +-7.07$ V. But after the diode, the voltage is zero, until the voltage is above $\sim$0.6 V, and then the voltage is the 5 V RMS AC voltage — $\sim$0.6 V, so gets up to a max of $\sim$6.4 V, and has a min of 0 V. Note that because the diode has a voltage drop across it while current is flowing through it, it uses power! Given a voltage drop of 0.6 V (which is independent of the current through it!) and a current of $I = V/R = (5$ V $- 0.6$ V$)/1$ $k\Omega = \sim$4.4 mA RMS, it would use $P = VI = 0.6$ V $* 4.4$ mA $= 2.64$ mW, when there is current flowing through it.[2]

---

[2]This is a rough calculation, because the waveform is more complicated, given the diode behavior.

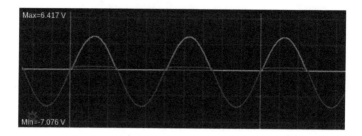

We now have a positive-only, or *rectified*, voltage, but it is fluctuating up and down at 60 Hz, the input 120 V AC wall frequency (in the USA). Transformers can not change the frequency of an AC voltage. To remove these fluctuations, what we need is a capacitor, as shown below. It will quickly charge up whenever there is an input voltage applied to it, so it will get up to $\sim$6.4 V. But then when the input voltage drops, it will supply voltage to the resistor load. The time for which the capacitor can supply a voltage to the resistor is given by the $R * C$ time constant for the particular R and C values. As long as $R * C$ is much larger than 1/60 seconds, the voltage will not drop much, since the capacitor will have plenty of charge to supply the current through the resistor until the input voltage is present again. For instance, in the example shown, we have a 100 uF capacitor and 1 $k\Omega$ resistor, so $R * C$ time constant of 0.1 s, which is 6 times larger than 1/60 s. That means it would take 6 cycles for the voltage to drop by $1 - \exp(-1) = \sim$63%, but the voltage is always getting topped up each cycle and starting over, so in fact it only has 1/6 of a "time constant to fall during each cycle", and decreases by only $1 - \exp(-1/6) = \sim$15%.

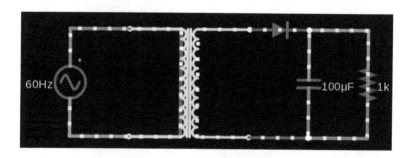

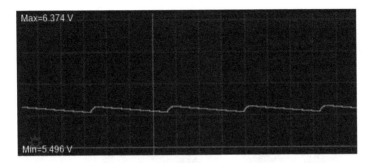

So far we have built what's called a *half-wave rectifier*, since it uses half of the AC sine wave, the positive part, and creates a DC output (although still with some "ripple"). But it would be even nicer to make use of the full sine wave, somehow charging the capacitor whenever the magnitude of the *negative* voltage is above the voltage of the capacitor. It would then only have to power the load resistor for half of a cycle, reducing the ripple. Such a circuit, cleverly constructed from four diodes, called the "full-wave rectifier", is shown below. Indeed the ripple has reduced to just ∼6%, using the same C and R.

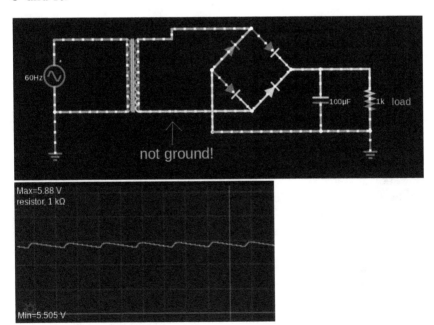

One thing to beware — the point after the transformer can not be connected to the bottom of the load resistor — they can not both be ground! (For the half-wave rectifier, they can be — try to see why.) Using the output of a transformer, this is not a problem, but in other cases it could be an issue. For instance, you can NOT use a full-wave rectifier on an AC signal that shares ground with the load, as shown below.

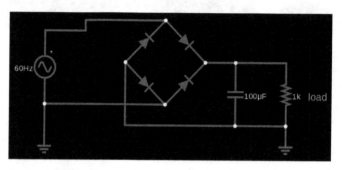

Voltage Regulation

After the full-wave rectifier, we now essentially have DC voltage. But there's still some ripple at 120 Hz (twice the 60 Hz input voltage). To get rid of that, we could use a bigger capacitor, but those start being large and expensive. There's also the issue of voltage *stability*. If our input voltage varies a bit (which it does — the power company may let it vary from ~110–130 V), or our transformer heats up and has a bit more resistance, etc, we'd still like to have the same output voltage. To solve all these issues, a good choice is to add a *voltage regulator* to the output of the capacitor, which then finally powers the load, as shown below. Voltage regulators use *active* components (transistors, which we'll see in the next chapter) to control and stabilize an output voltage. The input to the voltage regulator, V_{in}, must be a reasonably stable voltage, which is always at least $V_{in} > V_{out} + V_{drop}$, where V_{drop} is the *dropout voltage*. Some voltage regulators are adjustable, usually via some feedback resistors attached to an "adjustment" input, like in our example below, which uses an LM317. Other voltage regulators are even easier to use and have fixed output voltages, like the AMS1117, which has several versions each fixed to a typical output voltage: 1.5 V, 1.8 V, 2.5 V,

2.85 V, 3.3 V and 5.0 V. It claims to be a "low dropout" (LDO) voltage regulator, with a typical dropout voltage of only 1.1 V. These are reasonably complicated devices, and you must read the datasheet for the device you're going to use! In the end, we get a very stable output voltage, as seen below, where we compare the input voltage on the filter capacitor, to the voltage across the load resistor from the voltage regulator. It is 4.914 + −0.001 V, and should be stable over the long term and various temperatures to ~1%, according to the datasheet. Beware that voltage regulators do use power and thus can get hot! They usually have current limits below 1 A or so.

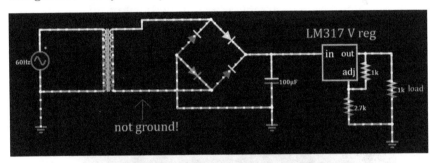

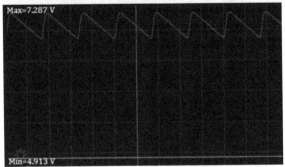

Switching Power Supplies

Finally, sometimes you have some particular DC voltage, and you need to create some other DC voltage — maybe larger, maybe smaller, maybe even a negative voltage. For this you might use a *switch-mode power supply*. The general idea is that the DC voltage is switched on and off, very rapidly, using a transistor (which we'll learn about next). The fraction of time the switch is on, called the *duty cycle*, controls how much voltage is passed to the output. An inductor and capacitor filter then smooth out the output voltage. For instance, the *buck converter*, shown below, is using a duty cycle of 25% at 100 kHz in order to output a smooth 1.934 V from a 5 V input.[3] You can also swap the inductor and diode in order to create negative voltages! To create larger voltages, you can use a boost converter. And it's even possible to combine the two into the buck-boost converter, which allows the output to be above or below the input voltage. Sometimes you may end up designing DC-DC converter circuits like this for your project, but more often these days you simply purchase premade modules, for less money than you could even buy the components on them!

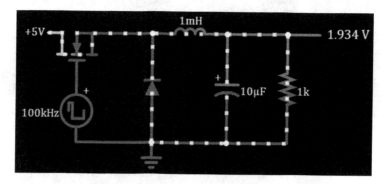

Lab: Inductance and Power Lab

Some things about inductors

1. Whereas capacitors are most constrained by their voltage limit ratings, for inductors you must pay most attention to the current rating. Since most inductors are made from loops of wire, and wire has resistance, it gets hot if you put too much current through it. Some inductors will also suffer (before burning up) from *saturation*, where the inductance decreases beyond some current limit. *Power inductors* are optimized for large currents.

---

[3]Ideally a duty cycle of 25% would give a voltage of $5 * 25/100 = 1.25$ V, but there are capacitive and inductive effects, as well as diode losses.

2. There are various types of inductors optimized for various properties, like accuracy and stability, high frequency behavior, low resistance, etc.

3. Most multimeters can not measure inductance directly, like they can for resistance or capacitance. To do that you need an *LCR meter*.

Exercises

1. Build the RLC bandpass filter circuit. Measure the frequency response of the filter, and compare it to simulations. Be careful to measure both V_{in} and V_{out}! V_{in} will not be what your signal generator says it should be — why?
   What is the resonant frequency? What does theory say it should be, for your component values?
   What is the Q factor? What does theory say it should be, for your component values?
   What is the phase shift of the filter at the resonant frequency? What about at higher or lower frequencies?
   Try all this for a few different component values, to test how resonant frequency and Q depend on those values.

2. Make a half-wave rectifier, using a 20 V peak-to-peak (V_{pp}) AC input signal from the signal generator (so we don't have to worry about using a 120 V transformer) and a 1 k resistive load. Compare the output voltage vs. time to simulations. What is the effect of the diode voltage drop? How much power is wasted in the diode? Add a filter capacitor to the rectifier output and compare the output to simulation. How big does C need to be to get a reasonably smooth output? Try a few different capacitor values.

3. Add a voltage regulator (LM7805) to your rectified signal to make a stable 5 V output. (Read the datasheet!)
   Compare input and output voltage *ripple*, with/without extra ceramic filter capacitors as suggested in the datasheet.

Chapter 4

Transistors

In the last chapter, we made use of diodes, which only allow current to flow in one direction. But how do they work? They start with silicon, or some other *semiconductor*, which is somewhere in between a conductor and an insulator. The trick is then to *dope* the silicon with slightly different atoms. If you add in some phosphorus atoms, for instance, they each have an extra valence electron, compared to silicon. This makes *N-type* silicon, which has some extra free electrons meandering around in it. If you add in some boron atoms, which each have one less electron than silicon, you get *P-type* silicon. It has some extra *holes* in it, places where electrons could be, basically positive charge carriers. Both these extra electrons or holes can conduct electricity pretty well, since they're free to move and have charge. So both *N-* and *P-type* silicon are decent conductors (though not great, and it depends on the amount of doping).

The magic happens when you put some *P-type* next to some *N-type* and let them touch along a surface. This is called a *PN-junction*, and this is what is inside a diode, as shown below. At the junction, electrons from the *N* side are able to jump over and occupy the holes in the *P* side. There is then a *depletion region* formed where there are very few free charge carriers, either electrons or holes, so it is not a good conductor. If you apply a voltage across the junction with a larger voltage on the *N* side than the *P* side, more electrons are drawn towards the higher voltage, and more holes are drawn towards the negative voltage, and the depletion region just grows larger. But if you do the reverse, and apply a voltage across the junction with a larger voltage on the P side than the N side, the depletion region shrinks. When the voltage is large enough, above the diode voltage drop of ~0.6 V, the depletion region is eliminated, and there are free charge carriers at each point in the junction, allowing current to flow through.

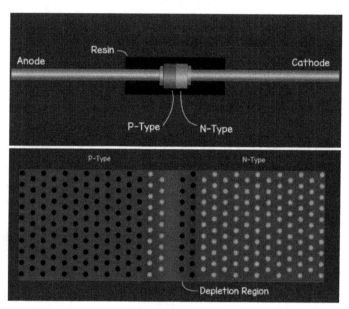

Transistors

A *transistor* is another semiconductor device, but with a third terminal, that will allow us to control the flow of current between the first two terminals. We'll first look at the *bipolar* transistor, which can be made by making an *N-P-N* sandwich, with a thin *P-type* layer between two pieces of *N-type* semiconductor. The *collector* and *emitter* are the first two terminals, attached to each of the *N-type* sides, and the *base* is the third terminal, attached to the P in the middle. The base to the emitter acts very much like a diode, so to get current to flow through the base, we need the base voltage (V_b) to be at least ~0.6 V larger than the voltage at the emitter (V_e). We also need the collector voltage (V_c) to be a little bigger than V_e, because current needs to flow from the collector to emitter. What's amazing is that we can then control the current flowing from the collector (I_c) and out of the emitter (I_e), by varying the current flowing into the base (I_b). Specifically, we find that, $I_c = \beta * I_b$, where β is the gain, or amplification factor, usually about ~100. So, by controlling a small current at the base, we can control a big current, ~100× larger, flowing through the collector and emitter! The current out of the emitter is the sum of the currents from the base and collector, $I_e = I_b + I_c$. We can put these together to see that $I_e = I_b + \beta * I_b = (1 + \beta) * I_b$.

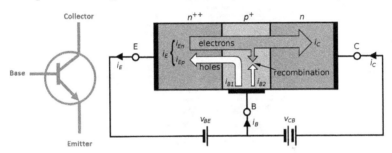

Transistor Switch

We'll start by using a transistor as a switch, as shown below, to turn on and off an LED. The largest the current could possibly be through the 300 Ω resistor is when the transistor is fully on, or *saturated*. The only resistance from 5 V to ground in that case is the 300 Ω resistor itself. The voltage across the 300 Ω resistor is $V_{Res} = 5\,V - V_{LED} - V_{CE}$, where V_{LED} is the voltage across the LED when it is on (about 1.8 V for this particular red LED), and V_{CE} is the voltage from the collector to the emitter when it is on (usually about 0.1 V). So when the LED is on, $V_{Res} = 5\,V - 1.8\,V - 0.1\,V = 3.1\,V$. That means the maximum current through the resistor, and thus also through the LED, since they're in series, is $I_{Res} = I_{LED} = V_{Res}/300\,\Omega = 3.1\,V/300\,\Omega = \sim 10\,mA$. This is also the largest possible current into the collector, I_C. Because the collector current is β times the base current, the maximum collector current will be reached as soon as the base current is 10 mA/100 = 0.1 mA. Thanks to the large gain of the transistor ($\beta = 100$), just a tiny current at the base and control the large current of the LED.

We can make this little 0.1 mA base current with a voltage source and a resistor $(10\,\text{k}\Omega)$. For a voltage source of 2 V, the voltage across the resistor would be $2 - 0.6\,\text{V} = 1.4\,\text{V}$, due to the voltage drop of the "diode" in the transistor from base to emitter. The base current would then be $I = V/R = 1.4\,\text{V}/10\,k\Omega = 0.14\,\text{mA}$, enough to fully turn on the transistor.

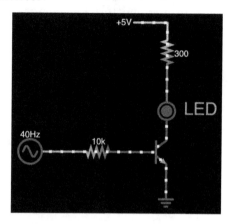

Transistors are "active" components — they use power! A tiny bit of power is used by the base current, which has to flow over the $\sim0.6\,\text{V}$ of voltage drop from base to emitter. (Usually a bit more is used by the base current resistor.) But most of the power is used by the $(100\times)$ larger collector current, which has to flow over the $\sim0.1\,\text{V}$ of voltage drop from the collector to the emitter. In our switch example, about 1 mW is used in the switch, whereas about 50 mW is being used by the LED's resistor and the LED.

You can also make a PNP bipolar transistor, which just reverses the collector and emitter. For PNP, the emitter is on top, at highest voltage, and the emitter to base current, through the emitter-base "diode" is what controls the emitter to collector current. It's a little bit unwieldy, since the transistor is on when the base voltage is small, and only turns off once the base voltage is large and near the emitter voltage. So sort of backwards. And PNP transistors tend to be more expensive and use more power, for technical reasons. But sometimes it is what's needed. For example, you can use it to switch the "high side" of the voltage in our switch, as shown in the following figure. The LED would be off whenever the voltage applied to the base resistor is above $5\,\text{V} - 0.6\,\text{V} = 4.4\,\text{V}$, and on below that voltage, approximately.

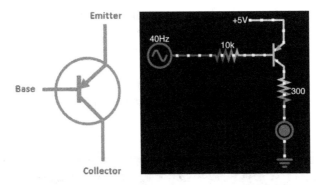

Transistor Emitter-follower

Let's now look at the following circuit, the dual supply emitter-follower. (*Dual supply* refers to the fact that there's both positive (5 V) and negative (-5 V) supplies. We'll see why this is useful soon.) The output voltage is just the same as the input voltage, minus the "diode" voltage drop of $\sim$0.6 V. So what's the point of the circuit then? The output *current* is β (actually, $\beta + 1$), about 100 times bigger! Remember that input impedance is given by the change in voltage divided by the change in current, at the input. If we change the input (base) voltage by ΔV, the emitter current will indeed change by $\Delta V/R$, just like for any resistor. But the change in the input (base) current is $\beta+1$ times smaller. So, the input impedance is not R, it's $R * (\beta + 1)$! Similarly, we'd find that the output impedance is $\beta + 1$ times smaller after the emitter follower.

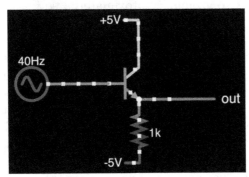

Being able to increase the input impedance and decrease the output impedance is extremely useful — we'll make frequent use of this later on, though usually not with a discrete transistor. This circuit still has some drawbacks. First, the output signal is always 0.6 V less than the input signal, thanks to the base to emitter voltage drop, as shown in the figure below. Relatedly, if the input voltage is smaller than $-5\,V + 0.6\,V = -4.4\,V$, the output voltage is stuck at $-5\,V$, since it can't go lower than the "negative supply" voltage. We'll see in the next chapter how the magical "operational amplifier" solves these woes.

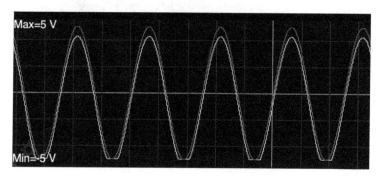

Transistor Amplifier

We can't really talk about transistors without showing how they can be used to amplify a signal. The following circuit is an example of a *common emitter amplifier*. The input signal (1 kHz, 0.5 V AC) is first put through a high pass filter (the 1 uF capacitor, combined with the 10 k resistor to ground, approximately), so $f_{-3dB} = 1/(2\pi RC) = \sim 16\,Hz$. The point of this is just to block the DC component, the voltage offset, of the input signal. The capacitor would be called a *blocking capacitor*. We want to do this so that we can adjust the *average* voltage at the transistor base, *bias* the input voltage, so that it's close to $\sim 1.3\,V$. This will keep the total base voltage swing, $V_B = 1.3 + -0.5\,V$, within the *active* range of the transistor. In the active range, the transistor is not purely on or off, it has a collector current which is β times the base current. That's what we want for our amplifier. If we're outside that range, we'll get clipping and distortion. (See what happens if you replace the bottom 10 k resistor with 50 k, for instance.)

With the transistor doing its job, in the active region, the collector current is β times the base current, which is proportional to V_B. The current through the collector resistor (10 k) and the emitter resistor (1 k), are about the same. (The emitter current just has a little extra from the small base current.) So the voltage drop across the top collector resistor is 10× larger than that of the bottom emitter resistor, since it is 10× larger resistance, the current is nearly equal, and $V = IR$. We have amplified the voltage by about 10×! This voltage drop is being *subtracted* from the +15 V supply voltage at the collector, so the collector voltage has ~10× the signal voltage fluctuation, but with opposite phase compared to the input signal. The last step is to again use a high pass filter to remove the offset DC voltage from the collector output, with the 1 uF and 1 MΩ, so we get a nice pure AC signal output.

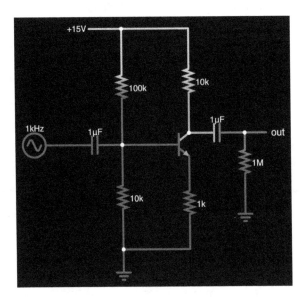

This amplifier has some serious drawbacks. First and foremost, it is complicated to bias the base correctly, and even then it only works for a small range of input signal voltages. It also is only an AC amplifier — the DC offset of the signal input and output are removed, which is needed for our biasing scheme. Lastly, the output signal is inverted (180 degrees out of phase) with the input. We'll see in the next chapter how operational amplifiers eliminate such issues (mostly).

Field-effect Transistors

So far we've been talking about bipolar transistors, which use a small base current to control the current from collector to emitter. Field-effect transistors (FET) are a different type of transistor that uses *voltage* (electric field), not current, to control the current from collector to emitter (often called *drain* and *source* for a FET). There is almost no base current, and very high base resistance — up to $10^{15}\,\Omega$! The most common type of FET is the Metal–oxide–semiconductor-FET (MOSFET), which uses a very thin layer of oxide to insulate the base (often called a *gate* for a FET) from the rest of the transistor. A Junction-FET (JFET) is like a MOSFET, but uses a different way of making the gate, a reverse PN-junction instead of an oxide barrier. They have lower gate resistance (*only* $10^{8}\,\Omega$ or so), but they are cheaper, less susceptible to ESD damage, and have some other improved performance properties. FETs also come in *n*-channel and *p*-channel types, just like bipolar transistors could be NPN or PNP.

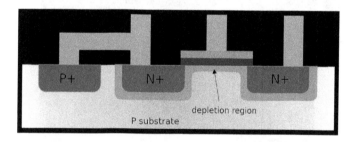

The FET switch is even simpler than the bipolar one, since you don't have to worry about base current, as shown in the following figure. The *n*-channel FET starts to turn on when the gate voltage is above the "gate threshold voltage" for gate to source, $V_{GS_{(th)}}$, usually about 2 V or so. (For a *p*-channel FET, it will start to turn off when the gate voltage is above $V_{GS_{(th)}}$.) It's not perfect, there is "on resistance" for current going from drain to source, about $\sim1\,\Omega$ to $\sim100\,\Omega$. The off resistance is very large.

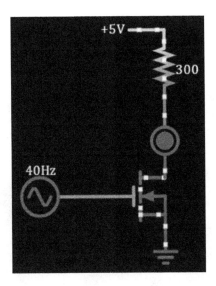

You can also make an *analog switch* from a FET, which allows a signal to pass through, or not, in either direction, as shown in the following figure.

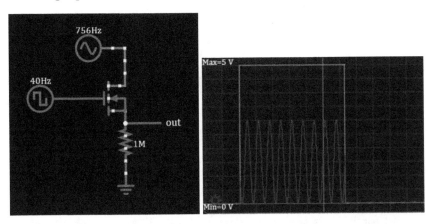

FETs are also perfect for digital logic, like this complementary MOSFET (CMOS) *inverter*, shown in the following figure, which uses two FETs, one *p*-channel and one *n*-channel. If the input is a high voltage, say 5 V, the top *p*-channel FET is off and the bottom *n*-channel FET is on, so the output voltage is 0 V. But if the input is a low voltage, say 0 V, the top *p*-channel FET is on, and the bottom *n*-channel FET is off, so the output voltage is 5 V. The input has been inverted! You can also make more complicated gates. From not-AND (NAND) gates, as shown in the following figure, which output a low voltage only if both of the inputs are high voltage, you can build *any digital logic*, even a full computer!

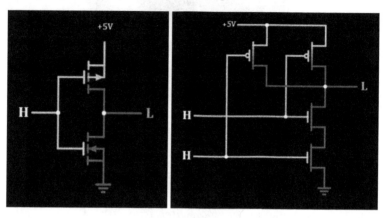

Lab: Transistors Lab

A few things about transistors

These are complicated components, and in general you really need to read the datasheet for any you use. But there are a few main things to look out for:

1. **Voltage rating**: for BJT this is the Collector-Emitter (or Collector-Base) limit and for FET this is the Drain-Source (or Drain-Gate) limit. These are usually at least 25 V, but can vary a lot.

2. **Current rating**: for BJT this is the Collector current limit and for FET is the Drain current limit. These are usually above 100 mA, but again can vary a lot.

3. Then there are some key performance stats.
   For BJT, the key stats are as follows:
   - the Base-Emitter saturation voltage, where the transistor turns "on"

- the Current Gain (aka: Beta or Hfe), how much the base current is amplified ... beware, this can vary a lot from transistor to transistor!

For FETs, the key stats are as follows:

- the Gate-Source Threshold Voltage, where the transistor turns "on"
- the Drain-Source On Resistance, the "resistance" of the transistor to current flow when it is fully "on"

4. You can test a transistor with a multimeter. Here's a decent guide: https://vetco.net/blogs/news/test-a-transistor-with-a-multimeter. You can test both the resistances between various terminals as well as voltage drops between terminals (to test the "diodes" inside). Many multimeters also allow you to measure the Hfe gain of a BJT transistor.

Exercises

1. Build a dual-supply emitter-follower, with an NPN like the 2N3904, and input an AC signal. Study the output and compare to simulation.

2. Build a common emitter amplifier and show how $V_{\text{out}}/V_{\text{in}}$ varies with gain resistor values. What is the DC base voltage?

3. Build a bipolar transistor "switch", and turn an LED on and off. Use a 1 Hz square wave as input to the base and observe the voltage across the LED.

   How does the base current compare to the load current?

4. Build an *n*-channel MOSFET switch, using a 2N7000, and turn an LED on and off. Use a 1 Hz square wave as input to the gate and observe the voltage across the LED. Make sure to exceed the gate threshold voltage at some point!

   What is the resistance of the MOSFET from drain to source when it's on?

   What is the gate current when it is on? off?

Does the circuit still work if you add in series the resistance of your body to the gate input? Why?

5. Build a CMOS inverter, using in addition a *p*-channel MOSFET. Input a square wave signal and observe the output voltage.

6. (Bonus) Build a CMOS NAND logic gate. Input two square waves with varying relative phase and observe the output voltage.

Chapter 5

Op-amps

Transistors were good, but they were very challenging to use (biasing, AC coupling, diode voltage drop, phase shift, etc.) and offered limited performance ($\sim 100\times$ gain, non-linearity, etc.). Plus their characteristics (gain, threshold voltage, etc.) vary from transistor to transistor a lot. In this chapter, we'll see how the *operational amplifier* (op-amp), combined with *negative feedback*, solves all these problems. The idea is to create a *differential* amplifier with very large gain, very large input impedance, and very small output impedance, and then use negative feedback to stabilize the circuit.

Differential Amplifier

A differential amplifier amplifies the difference between its two inputs and sends it to its output. The inputs are labelled "+" (the non-inverting input) and "−" (the inverting input), as shown below. There are also positive and negative voltage power supply connections (called *rails*) to the amplifier, say $+5$ and $-5\,\text{V}$, but they are often not shown explicitly in schematics. The output is equal to β (gain) times the voltage of the non-inverting input minus the voltage of the inverting input. But β is very large for an op-amp, around 100,000 (or 100 dB). The output can not be larger than the positive rail or less than the negative rail though. So, if + is just 1 mV higher than −, the output is equal to β times $1\,\text{mV}$, or

$100000 * 1\,\text{mV} = 100\,\text{V}$, but the maximum (equal to the positive rail) is just $5\,\text{V}$, so the output is $5\,\text{V}$. If $-$ is $1\,\text{mV}$ higher than $+$, then the output would be $-100\,\text{V}$, but again the minimum (equal to the negative rail) is $-5\,\text{V}$, so the output is $-5\,\text{V}$. Because of the huge amplifier gain, the op-amp appears to be very unstable. A tiny change in an input, when the inputs are near equal, causes a seismic shift in the output to either rail.

This behavior, where the output goes to either the high or low rail depending on whether the $+$ input is above or below the $-$ input, is useful as a *comparator*. Sometimes you want to know when a signal is above or below some given *threshold* voltage. An op-amp can be a sensitive comparator by inputting the threshold to the $-$ input and the signal to the $+$ input. You can also buy specialized comparator chips which are optimized for this task.

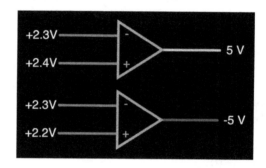

Negative Feedback

The key to taming this behavior is *negative feedback*. Here's an op-amp version of the emitter-follower, often called a *buffer*, shown below. The output is connected, with a wire, back to the inverting input. The $+$ and $-$ inputs to op-amps have very high impedance, usually by using FET transistors where the input goes into the gate, so they draw (almost) no current. What happens as we change the voltage at the non-inverting input? If the voltage at the inverting input is higher than that at the non-inverting input (by even a mV!), the output is forced towards the negative rail. But since the output is connected back to the inverting input, the voltage at the inverting input will also be driven towards the negative rail. This continues until the voltage at the inverting input is no longer higher than that at the non-inverting input, so until it is equal to that at the non-inverting input. The reverse would also happen if the voltage at the inverting input were lower than that at the non-inverting input. The output, and thus the inverting input, would get forced towards the positive rail until it is equal to the non-inverting input. This is negative feedback. If the output is too high, it is forced lower. If the output is too low, it is forced higher. When there is a negative feedback loop like this, the output tries to make the two inputs have the same voltage.

The end result is that the output will always equal the voltage at the non-inverting input, as long as it is within the voltage limits the op-amp can output. If the non-inverting input is larger than the maximum voltage the op-amp can output, it will output that maximum. If the non-inverting input is smaller than the minimum voltage the op-amp can output, it will output that minimum. This is an improvement on the emitter-follower we built with a transistor. There is no more diode voltage drop from the input to the output. And the output impedance is not just β times lower than the input's impedance, it is the output impedance of the op-amp, which is just a $\sim$ few Ω or so. The input impedance is also very large, in the MΩ's to even GΩ's or so. It even uses much less power than the single transistor version, thanks to clever internal optimizations. In case you're curious how all this is possible, the simplified schematic of the popular LM358B op-amp (which sells for just a few cents!) is shown in the following figure.

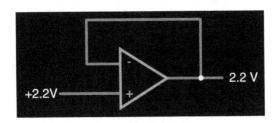

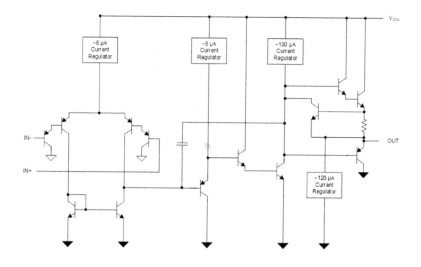

Amplifier

We can easily make the gain greater than 1, so our op-amp actually amplifies, as shown in the following figure. (Note that here we've drawn the non-inverting input on the top! Sometimes, as is the case here, doing so makes the diagram prettier.) How does this work? Remember that the output must change so as to keep the input it's connected to (the inverting input in this case) equal in voltage to the other input, when there's negative feedback at work, as is the case here. Let's say the non-inverting input is at $1\,\mathrm{V}$. What voltage does the output have to get to such that the voltage at the inverting input is also $1\,\mathrm{V}$? There's a voltage divider present, with the output at the top, the inverting input at the middle, and ground at the bottom. So, the output must get to $3\,\mathrm{V}$ such that there's $1\,\mathrm{V}$ present at the inverting input, thus the gain of the amplifier is 3. Whatever voltage is present at the non-inverting input (positive or negative!), the output voltage will be 3 times larger. In general, the gain is $1 + R_2/R_1$. Note that in the limit that R_2 is $0\,\Omega$ (a wire) and R_1 is infinity Ω (open), we have gain of 1, so we recover the simple emitter-follower buffer from above. The input impedance is very large, which is a great thing, since it goes directly into an op-amp input. But, there's no way to get a gain less than 1, or a negative (inverting) gain. And the *common mode* voltage (average of the two inputs) varies far from 0, leading to possible distortion.

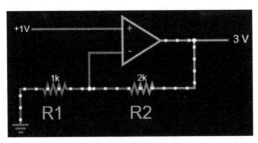

Inverting Amplifier

The inverting op-amp configuration, shown in the following figure, solves these issues. The tradeoff is that input impedance is now small, equal to R_1, and it always inverts the signal. The non-inverting input is grounded. So the op-amp must adjust the output so that the inverting input is also at $0\,\text{V}$. The inverting input is called a *virtual ground* in this case. It's not really a ground, but the voltage there will be kept at $0\,\text{V}$ by the op-amp (or at least it will try to keep it at $0\,\text{V}$). There is again a voltage divider, with the output at the top, the inverting input in the middle, and the input signal at the bottom. For the case shown in the following figure, if the input signal from the signal generator is at $1\,\text{V}$, the output would need to be at $-3\,\text{V}$ to make the voltage at the inverting input $0\,\text{V}$. Thus, both inputs are always very near ground, so the common mode voltage is very small, thus there's very little risk of distortion. In general the gain is $-R_2/R_1$, where the "$-$" means that the output signal is inverted compared to the input.

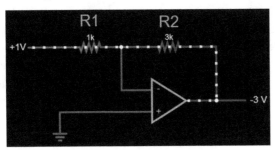

Signal Adders

Another important use of op-amps is adding signals together. For instance, we can add another signal to the input of the inverting op-amp configuration, as shown in the following figure. The output must then be the negative of the sum of the input voltages at the inverting input, since the inverting input is a virtual ground. You can easily make a *weighted sum* by changing the values of the inputs' resistors. The gain for each of the inputs is R_2/R_1. So, you can also make the gain different from 1 overall, either larger or smaller. In general, $V_{\text{out}} = -R_F * (V_1/R_1 + V_2/R_2 + V_3/R_3 + \cdots)$. The drawbacks are the same as those of the inverting amplifier above, namely small input impedance and the fact that the output is inverted.

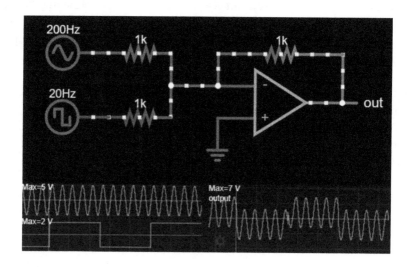

We can also add another input signal to the non-inverting op-amp configuration, as shown in the following figure. In this case though, we have to be careful to keep the gain of the op-amp feedback equal to the number of inputs. So for two inputs, we would want a gain of 2, and since the gain is $1 + R_2/R_1$, we would want R_2 equal to R_1. If you want to *average* the inputs, you would just make the gain 1, i.e. use a buffer.

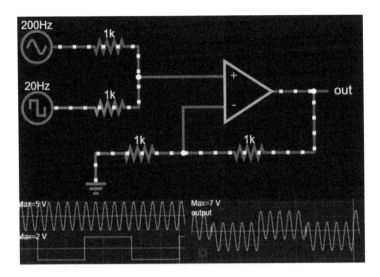

Limitations

Op-amps are not perfect, of course, and have many limitations to be aware of. Before using any complicated *integrated circuit* (IC) like this, you must read the datasheet! It will tell you nearly everything you could want to know about the part. Let's have a look at the datasheet for the LM358B op-amp, which we showed the simplified schematic of above, to get an idea of what the performance parameters and limitations of op-amps are.

The supply voltage (positive rail — negative rail) is limited to between 3 and 36 V. So, you could set the positive rail to anywhere from 3 to 36 V, as long as the negative rail is ground (0 V). That would be single supply operation. Or you could use it with split supplies, from ±1.5 to ±18 V, for instance 5 V on the positive rail and −5 V on the negative rail. You can get other op-amps that run at voltages down to ∼1 V, and others up to ∼400 V, but they must be specialized for such ranges.

The part will work fine in temperatures from −40 to 85°C, which is −40 to 185°F, easy to remember! This is standard for "commercial" grade parts. For wider temperature ranges you can get "industrial", "automotive", or "aerospace" grade parts.

Next we see in "electrical characteristics" many important parameters. Note that these are specified only under certain conditions, here for 5–36 V of supply voltage, at 25°C (room temperature), inputs and output midway between the positive and negative rails (e.g. ground for ±5 V supply voltages), and a 10 k output load connected to that same voltage. The *input offset* voltage is the difference in the + and − inputs' voltages for when the output is midway between the positive and negative rails. In other words, if the rails are ±5 V, and the output is halfway between, at ground, so the op-amp "thinks" the two inputs are at the same voltage, but the inputs are actually 1 mV different from each other, then the input offset voltage is 1 mV. For the LM358B, the input offset voltage is specified to be ±0.3 mV, meaning you don't know whether the inverting or non-inverting input would actually be higher voltage by that amount, when the output is balanced. This is an important parameter in some cases, for instance when using the op-amp as a comparator, it would limit how accurate the comparator would be compared to the input threshold voltage. The input offset voltage is also temperature dependent, and for this op-amp varies by ±3.5 uV for each °C.

The *power supply rejection ratio* (PSRR), tells you how much noise from the power supply rails will leak through into the output. For this op-amp it claims it is just 2 uV/V, meaning if your power supply rail had 1 V RMS of noise on it, the output would only get an additional 2 uV RMS of noise on it. But this is misleading! The PSRR is highly dependent on frequency, and is also different for noise on the positive vs negative rails, as shown in a plot later on in the datasheet. At a frequency of 10 k, the PSRR for the negative rail is only ~40 dB, or 10 mV/V!

The *common mode voltage*, the average of the + and − inputs, can vary from the negative rail up to 2 V from the positive rail. Ideally, the op-amp output behavior would not be affected by the common mode voltage. The output should just depend on the difference between the inputs. But many op-amps have limited common mode voltage range, which is one reason the inverting op-amp configuration (which always keeps the common mode voltage to 0 V) is sometimes preferred to the non-inverting configuration.

The *common mode rejection ratio* is like the PSRR, defining how simultaneous noise on both inputs finds its way into the output. Ideally the op-amp output would not be sensitive to any common mode voltage variations or noise, and only depend on the voltage difference between the outputs.

Input current and impedance are specified, but these are usually negligible. Ideally, the op-amp should have infinite impedance at its inputs, and real op-amps do have very large impedance, here $10\,M\Omega$ for the LM358B.

The op-amp *input voltage noise*, from just the op-amp itself, is specified in two ways. The first is in the frequency range from 0.1 to $10\,Hz$ (so very low frequency), here $3\,uV$ peak-to-peak. Note that this is input noise, so if the op-amp is configured to have a gain of $100\times$, the output noise would be $100\times$ larger than this, or $0.3\,mV$. It is also specified as a *noise density* for higher frequencies, here $40\,nV/\sqrt{Hz}$, at $1\,kHz$. This means that in the frequency range of $1\,kHz$ to $10\,kHz$, which is $100\sqrt{Hz}$ (because $\sqrt{10\,k}$ is 100), there would be $40\,nV * 100 = 4\,uV$ of noise at the inputs. The larger the frequency range, the larger the noise (increasing like the sqrt of the frequency range). This is because noise at high frequency is usually dominated by ("white") thermal noise, which is flat in power vs. frequency, but voltage is proportional to the square root of power.

Gain

The op-amp gain is specified in the open-loop (no feedback) case as $140\,V/mV$. (Note that it could be as low as $70\,V/mV$, not all LM358B will be the same! This is true for many parameters in datasheets — there is a typical value, but also a min and/or max it could be.) Op-amps have huge gain, as we said — ideally infinite.

But more useful practically are the gain in the case that feedback is being employed. The *gain bandwidth product* (GBW) is a limitation on the product of the bandwidth (how high in frequency the op-amp will work up to) and the gain. For example, for the LM358B, the GBW is $1.2\,MHz$. So for a gain of 1, the bandwidth will be $1.2\,MHz$, meaning the gain of the op-amp will be $-3\,dB$ at $1.2\,MHz$, or about 0.7. It is a low-pass filter with a cutoff frequency of $1.2\,MHz$. But if the gain is set to 10, the bandwidth will now be just $120\,kHz$, because the product of the gain and bandwidth is constrained to be $1.2\,MHz$. So at a gain of 10, the op-amp will act as a low-pass filter with a cutoff frequency of $120\,kHz$. Why does the bandwidth decrease as the gain increases? Well higher gain requires larger feedback resistance, so for fixed input capacitance the $R * C$ is larger, and the output takes longer to change voltage. (Of course it's more complicated than this, but that's the basic idea.) Beware that GBW is just a good "rule of thumb" that tends to describe op-amp behavior. For details about the actual bandwidth achievable at various gains, read the datasheet carefully, including the detailed plots of bandwidth at various gains. Even the values of the feedback resistor (for the same R_2/R_1 and thus same gain) can affect bandwidth. The best bandwidth usually is obtained with an R_2 of a few hundred Ohm or so, but again, read the datasheet. Also beware that bandwidth is lower (by up to a factor of 2, for the gain of -1) for the inverting amplifier, since more "open-loop gain" is needed (for the same gain with feedback) in this configuration.

Slew-rate

The output voltage is also limited in how fast it can change in absolute terms, not just in frequency. This is called the *slew rate*, which for the LM358B is 0.5 V/us. In other words, even if it had infinite bandwidth, if the amplifier is outputting a square wave, the maximum slope of the wave when it is going from low to high (or high to low) is 0.5 V/us. So, if you're asking for the op-amp to output a sine wave with a frequency of $f = 1\,\text{MHz}$, and amplitude of 1 V (peak), $\sin(2\pi ft)$, the slope of it at each moment t is $2\pi f * \cos(2\pi ft)$, and cos has a maximum of 1 V, so the maximum slope is $2\pi f$. That's 6.28 V/us, about 10× higher than the slew rate of the amplifier, so you would have serious distortion of the sine wave output. You'd have to reduce the amplitude of the output by a factor of 10 to remove the distortion. So, at large voltage output swings, large output signals, you become limited by slew rate. At lower output amplitudes, you're more limited by bandwidth, in general.

Voltage Output Range

Many op-amps can not actually output a voltage as large as the positive rail or as small as the negative rail. They might get within

1 V of them or so. Or maybe within 1 V of the positive rail and all the way down to the negative rail. Beware — and read the datasheet. Op-amps that can get their output to the rails are called rail-to-rail op-amps. Some op-amps, called single supply, are optimized for running from a single positive rail, like 5 V, and ground. Others, called split supply, are meant for running from a + and − rail, like ±5 V. Op amps also have limited output current ability, usually in the 20–200 mA range.

Stability

Perhaps the biggest issue with op amps though, is they can oscillate. We relied on negative feedback to make the op-amp output stable. If the output was too large, the negative feedback drove the output smaller. But recall that when capacitance and/or inductance are included, the output signal phase can be altered, or it can be delayed, compared to the phase of the input signal. If the output phase for some input frequency is changed by more than 180 degrees, but the gain of the feedback is still above 1, then there is not negative feedback anymore — there's positive feedback! That means the amplitude of that output frequency will increase to the maximum it can, which is very bad. Op-amp datasheets will almost always have a plot like the one below, where the phase and gain are plotted together vs frequency. It is critical that the phase (of the feedback signal relative to the input) remains above 0 whenever the gain is above 1 (0 db). In this case, the phase is about 45 degrees still when the gain falls below 0 db, so this op amp has plenty of phase margin. But adding additional capacitance or inductance in the feedback loop can decrease the phase margin and lead to oscillation. If you find oscillations happening, try to decrease capacitance and inductance in the feedback loop. Use a PCB (discussed later), not a breadboard, and the smallest components possible. Use smaller feedback resistance values, e.g. 50–500 Ω is a good range. If all else fails, you may need to compensate the feedback loop, by adding more capacitance in parallel with the feedback resistor, maybe 1–10 pF.

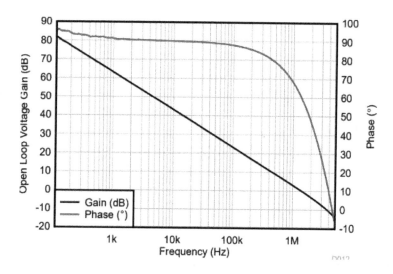

For these reasons, op amps are usually not good at driving capacitive loads. For instance, the LM358B specifies a limit of 100 pF that can be attached to the output. Any more is likely to cause oscillations of the output or other problems. The solution for driving capacitive loads is usually simple — just place a resistor (10–100 Ω or so usually does it) between the output and the capacitance.

Lab: Op-amps Lab

Some things about op-amps

1. Start with one like the LF411. Look at the pinout and specs on the datasheet. Read what each pin is used for. Some can be left *floating*, i.e. disconnected, like pins 1, 5, and 8 on the LF411.

2. Some op-amp packages have two op-amps in them. They're called *dual op-amps*. An example is the LMC662 — see the datasheet. The V+ and V− rails are shared by the two op-amps.

3. For integrated circuits (IC's) like op-amps, make sure you know where pin 1 is! It is usually indicated by a dot on the package.

4. You may need to *decouple* the power rails (add 0.1 uF capacitors between each rail and ground) to prevent noise and/or oscillations.

Exercises

1. Build a simple op-amp "open-loop" (no feedback) circuit: See what happens to the output when one input is above or below the other input. Can you ever get the output to be not "saturated" (equal to one of the voltage rails)? What is the difference in voltage between the two inputs when the output switches from high to low, or low to high? This voltage difference is the *input offset voltage*. Do they agree with the datasheet?

2. Build the op-amp emitter-follower buffer. Verify that an input signal is reproduced by the output.

   What changes if you put a 1 MΩ resistor in series with the input signal? Is the output different? What does that say about the input impedance of the op-amp? Does it matter if you use 1× or 10× mode for your scope probe? Why?

   Check that the output impedance is really small, by placing smaller and smaller load resistors on the output and measuring the output voltage. Beware that the *current* limit of the op-amp output is only ~20 mA, so keep the output voltage small when testing small loads!

3. Build the non-inverting amplifier. Input a 100 Hz sine wave of various magnitudes. What is the maximum output "swing" (how high and how low can the amp output be)? What is the gain, for various R_1 and R_2? How high in frequency can you go before you see distortion of the output waveform and −3 dB of the output voltage rms? How does that relate to the GBW product for the op-amp?

4. Build the inverting amplifier. Input a 100 Hz sine wave of various magnitudes. What is the gain, for various R_1 and R_2? Measure the input impedance of the amplifier by adding additional input series resistors and observing the output.

5. Build the summing amplifier and add two signals. What is its output voltage gain? Try making one of the inputs a DC voltage.

6. (Bonus) Build a "perfect" full-wave rectifier (no diode voltage drop, and can work with a real ground): https://www.falstad. com/circuit/e-amp-fullrect.html

7. (Bonus) Build the simple op-amp follower buffer again and measure the "slew-rate" of the op-amp. Input a square wave and look at the edges in detail with the scope. Compare for instance the edge rise time at the op-amp's input to that at the op-amp's output. Does it agree with the datasheet?

   Measure slew rate a second way by inputting a large sine wave and see at what frequency the wave begins to distort. Are the two measurements of slew rates compatible?

   How do the limitations from GBW product and slew rate compare? Which one is dominant first for small signals? For large signals?

Chapter 6

Signals and Cables

Our electronic devices are usually designed to take input *signals* of interest, process them in some way, and output signals to other places. For instance, a device could take in two signals from a signal generator, sum them using an op-amp, and output the summed signal to an oscilloscope. A "signal" is usually going to mean a voltage which is some function of time, $V(t)$. Up to now, we've taken for granted that the $V(t)$ at the output of the signal generator was passed immediately and flawlessly to the input of our device. It was assumed to be unchanged after going through the wire or cable, and we assumed there was no delay or time offset between one end of the cable and the other. Similarly, we assumed that signals were passed flawlessly within our device, and from the output of the device arrived immediately and faithfully at the input to the oscilloscope. This was a pretty good assumption and approximation for the setups we have considered up to now, but it can be completely wrong in other situations! It's not just that our cables and devices are "imperfect" — there would be issues even if you made them out of ideal components and zero-resistance wire. The laws of E&M add unavoidable complications we must consider carefully in certain situations.

Frequency and Bandwidth

Let's imagine we have a signal generator being output to our device 3 m away, using a cable. We have one wire going from the output of the signal generator to an input of our circuit, and we have another wire going from the ground of the signal generator to the ground of our circuit, as shown in the Figure. We'll be nice for now and even assume the wires have zero resistance. Sometimes we forget about this ground wire, but it must be there for us to have a DC current flow, it will become critical for some issues in this chapter! Note that even if you didn't hook up this ground wire, as long as the device was "plugged in" (not being powered by a battery or isolated with a transformer etc.) current would return from the device into the ground connection of the device, into the electrical wiring of the wall, and up through the ground connection of the signal generator to complete the circuit. This would just be a longer and more circuitous ground wire, and would cause *more* problems, as we'll see later.

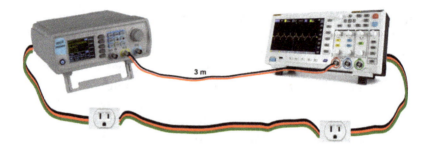

Have our signal generator start at $0\,\mathrm{V}$, at time $0\,\mathrm{s}$, and then go up to $1\,\mathrm{V}$ for $10\,\mathrm{ns}$, and then turn back off, as shown in the following figure. This is an ideal square pulse, $f(t) = 1$ for the range of time $[-T,\ T]$ and 0 elsewhere, with $T = 5\,\mathrm{ns}$. (We'll shift it so that it's symmetric around $t = 0$ to make the math easier.) What are the frequencies of this signal? Any signal that is not a steady sine wave of a single frequency is a mix of frequencies. If we did a Fourier transform of this ideal square pulse we would find:

$$F(\omega) = \int_{-T}^{T} e^{-i\omega t} dt = \frac{-1}{i\omega}\left(e^{-i\omega t} - e^{i\omega T}\right) = \frac{2\sin(\omega T)}{\omega}$$

This is also known as the sinc function for you mathematicians out there. Here's what that looks like, in the "time-domain" on the left, and in the "frequency-domain" on the right:

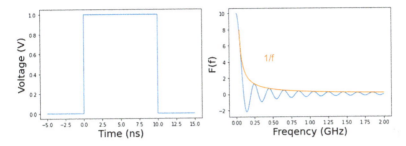

Much of the power is contained within frequencies below $1/10\,\text{ns} = 100\,\text{MHz}$ for our 10 ns pulse. That makes sense, since the fundamental frequency of a wave is 1/period of the wave. But note that the frequencies you need to make the sharp square pulse also extend all the way out to infinity, and only decrease in power like $1/f$. A square pulse is really built out of a sum of an infinite number of (odd) harmonics, with each harmonic, n, being n times smaller than the fundamental. We can see this in the following figure where we plot the sum of the first N (odd) harmonics, for $N = 1, 3, 5, 10, 20$, and 100. For example, for $N = 3$ it is $\sin(x) + \sin(3x)/3 + \sin(5x)/5$, times $2/\pi$ for normalization.

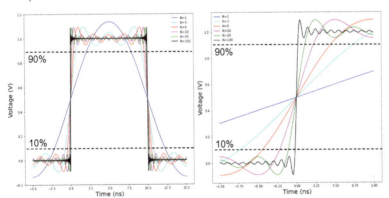

By the time you get to including the 5th (odd) harmonic, you have a reasonable looking square wave. But that would be at a frequency 9 times that of the fundamental, since the 5th (odd) harmonic is $\sin(9x)/9$. That means we would need to include frequencies up to about 1 GHz for our 10 ns pulse example to get a "reasonable" square pulse, and all the way up to 20 GHz ($N = 100$) to get a "really good looking" square pulse! By the way, you may notice that the approximations always overshoot (and undershoot) the true pulse, by about 10%. This even happens for large N — it always overshoots by the same amount, just the width of the overshoot in time gets smaller! This is called the *Gibbs phenomenon*, and you will see it when looking at pulses on the oscilloscope, due to bandwidth limitations.

A way to quantify the sharpness of the square pulse is by measuring the *rise time* (or *fall time*) of the pulse edge. Generally this is defined as the length of time it takes to get from 10 to 90% of the signal height. You can see clearly that as we include higher frequencies, the rise time gets smaller. This is because the slope of $\sin(nx)$ is at most n, since $\frac{d}{dx}\sin(nx) = n\cos(nx)$, and $|\cos(nx)| \leq 1$. To get a small rise time, we need a large slope, so we need a large n, thus a high frequency harmonic.

The reason we need to consider signals in the frequency domain is because *every* electronic device and component has some limit to how large a frequency it can handle. This is because there is always a limit to how fast a voltage can change. There is always some capacitance at the output of the signal generator, between the signal output and the ground terminal. Adding charge to the signal output will inevitably create an electric field between it and the ground terminal (stray capacitance). Since $I = C\ dV/dt$, an infinitely rapidly changing voltage would require infinite current to charge that capacitor, which is impossible because the output also has impedance (inductive and resistive). Let's say the output capacitance is 10 pF and output impedance is 10 Ω. There's effectively an RC filter at the output, with a cutoff frequency $(-3\,\mathrm{dB})$ of $1/(2\pi RC) = {\sim}1.6\,\mathrm{GHz}$. We say that the *bandwidth* of the device is thus 1.6 GHz. That doesn't mean it includes no frequencies above 1.6 GHz, just that the frequencies start to be significantly decreased in amplitude above that frequency:

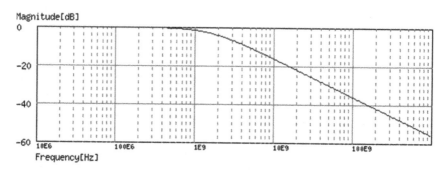

Since the higher frequencies are being diminished, the rise time (and fall time) of the square wave can not be too small. A good rule of thumb is that the -3 dB bandwidth of a signal $\approx 0.35/$risetime (or falltime). So, in our case with a -3 dB bandwidth of 1.6 GHz, we would have a rise and fall time of about $0.35/1.6$ GHz ≈ 0.2 ns. Our square pulse would look roughly like the $N = 20$ curve in the Figure above, since frequencies above $2 * 20 * 100$ MHz $= 4$ GHz are starting to be really attenuated. Remember that -6 dB is a factor of 2, so by $2 * 1.6$ GHz $= 3.2$ GHz the signal is diminished by a factor of 2. The $N = 20$ curve does indeed rise from 10 to 90% in height in about 0.2 ns.

Every electronic device, such as a signal generator, amplifier, multimeter, or oscilloscope, has a limited bandwidth. A multimeter might have a bandwidth of 10 kHz for instance. That means the RMS voltage measured by the device at 10 kHz will be 3 dB smaller ($\sqrt{2}$, or about 1.4 times smaller) than at low frequency. And it will fall by another 3 dB an additional factor of 2 smaller each additional doubling of the frequency (e.g. -12 dB, or 4x smaller at 40 kHz). You'll sometimes hear this said as -6 dB/octave, so 2 times smaller, at 20 kHz, and so on. A typical oscilloscope may have a -3 dB bandwidth of 50 MHz (depending on the settings and how the probes are being used!). For enough money ($\sim$\$1 M!), you can get an oscilloscope with bandwidth of $\sim$100 GHz today.

The Need for Speed

Once we start to play with higher frequency signals, many issues of signal integrity and noise become important. So, if higher frequency signals are so difficult to deal with, why use them? Sometimes you may have no choice, for instance maybe you need to be receiving or transmitting that frequency from an antenna. Wi-Fi works at 2.4 or 5 GHz. Or maybe you are detecting a signal that changes very fast, such as from a photo-multiplier tube. But even more commonly these days, we are regularly forced to deal with higher frequency signals because we are using or making devices that are trying to transmit, receive, and process information quickly — usually digital information. Even a \$5 ESP32 has a 240 MHz processor, and its inputs and outputs go from voltages of 0 to 3.3 V in less than $\sim$1 ns. Recall that for a 1 ns risetime we have to care about frequencies up to at least 350 MHz, more likely ~1 GHz for a nice looking pulse. Imagine we want to send 1 billion *bits* of information per second (1 Gbps — about the speed of a single channel of USB3). A bit is the smallest unit of *information*, a yes or no, a 1 or 0, an on or off. That means we need to be able to turn a voltage from high to low (or vice versa) at least once each ns, meaning each transition's fall (or rise) time must be significantly less than 1 ns, maybe ~200 ps, since the signal should also remain in the high or low state for a reasonable amount of time. We thus need to care about frequencies up to ~5 GHz by our rule of thumb above! Now you will likely not be developing a PCIe-6 interface which has ~10 ps rise/fall times after reading this book alone, but you can appreciate how even modest modern digital electronics routinely forces us into the higher frequency world these days.

Transmission Lines and Termination

Let's now consider what we would see at the input to the oscilloscope after creating our 10 ns long "square" pulse at the output of the signal generator. For one thing, the voltage at the input can not start rising until at least a time $t = L/c$ has passed, where L is the distance from the signal generator to our device input, 3 m, also assumed to be the length of the signal wire here, and c is the speed of light (in empty space). This is a basic requirement of causality baked into special relativity and at the heart of Maxwell's equations of E&M. No signal can travel faster than the speed of light. So, there will be a time delay between the output of the signal generator and the input to our device, given by how fast the signal *propagates* down the wire(s) and through space, which must not be faster than c.

To see why this is, consider our pair of parallel wires in more detail. Each little length of the wires (dx) have some resistance (R), some inductance (L), some capacitance between them (C), and some conductance between them (G). They form a *transmission line*:

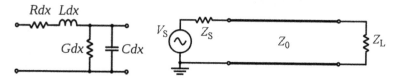

The general equations for the voltage and current on these lines, as a function of position (x) and time (t) are called the *telegrapher's equations*:

$$\frac{\partial}{\partial x}V(x,t) = -L\frac{\partial}{\partial t}I(x,t) - RI(x,t)$$

$$\frac{\partial}{\partial x}I(x,t) = -C\frac{\partial}{\partial t}V(x,t) - GV(x,t)$$

which can be combined to a second-order differential equation for the voltage (and a similar one for the current):

$$\frac{\partial^2}{\partial x^2}V(x,t) - LC\frac{\partial^2}{\partial t^2}V(x,t) = (RC + GL)\frac{\partial}{\partial t}V(x,t) + GRV(x,t)$$

In the limiting case that the R and G are small compared to L and C, which is normally a pretty good approximation, you get a simple wave equation for the voltage (and a similar one for the current):

$$\frac{\partial^2 V}{\partial t^2} - u^2\frac{\partial^2 V}{\partial x^2} = 0,$$

where $u = 1/\sqrt{LC}$ is the speed of the wave propagation down the line. For the wires in our case, which are just two parallel wires, with no dielectric between them, the speed will be close to c, the speed of light in vacuum. Details of common cable types will be discussed below.

The transmission line will be driven with some source impedance, Z_s, which is the output impedance of the signal generator, for instance. The transmission line itself will also have a *characteristic impedance* (for small R and G still), $Z_0 = \sqrt{L/C}$, which follows from the relation between I and V on the line, $V = IZ_0$. Both the L and C per unit length will depend on the geometry of our wires, e.g. how thick they are, how far apart they are, dielectric between them, etc. For two 1 mm diameter wires, 5 mm apart, in air, the Z_0 is about 300 Ω, for instance. The line will lastly be *terminated* with some load impedance, Z_L, which in our example is the input impedance of our oscilloscope.

Note that the current begins to flow *immediately* when the signal is launched into the line, in both the signal wire (towards the other end of the line) *and* the ground wire (back towards the source). The current in these wires is completed by the *displacement current* flowing between the signal conductor and the return conductor. Here's a nice simulation of the ladder network. Veritasium also made a couple great videos about the subject!

When the signal does reach the end of the cable, at the input to the oscilloscope, it will not just simply be "absorbed" at the input of our device, unless we are careful. We want the displacement current that was flowing between the signal and ground to be smoothly replaced by the charge current through the load at the end of the line, i.e. the old displacement current should equal the charge current in the load, or $Z_L = Z_0$. If the impedance of the input to our device is too large, $Z_L > Z_0$, charge will build up there, and a partial reflected pulse will be sent back down the cable towards the signal generator! If the impedance of the input is too low, $Z_L < Z_0$, too much current will flow, and an inverted partial reflected pulse will occur. To avoid signal reflections, we'll see that the input of the device must be *impedance matched* to the wires or cable, $Z_L = Z_0$. These reflections can even keep going back and forth (until they are eventually damped by the resistive effects of the cable in real life) if the impedance at the output of the signal generator is also mismatched, $Z_S \neq Z_0$. If $Z_S = Z_0$, the reflected pulse from the load end will get absorbed into the source and not re-reflected. Very often you'll arrange things so that *both* the source and load termination impedances match Z_0, so that you have the least chance of reflection problems, even if you mess up either Z_S or Z_L a bit.

In general, the fraction of the pulse amplitude that is reflected is $\Gamma = \frac{Z_L - Z_0}{Z_L + Z_0}$, and the same for Z_S and Z_0. The power reflected is the amplitude squared, so if the impedance mismatch is 10%, the reflected power is only 1%, not too bad. In general, all these three impedances can be complex, so there will be a phase shift of the reflected pulses as well, but usually we use nearly pure resistive source and load impedances (small capacitance and inductance) and have nearly completely real line impedance (from small R and G). Note that in the limiting cases of $Z_L \gg Z_0$ (an open end) you get 1, a reflected pulse, and for $Z_L \ll Z_0$ (a shorted end) you get -1, a reflected inverted pulse (so if at first the signal wire were 1 V and the ground wire 0 V, now the signal wire will be -1 V compared to the ground wire). Whatever voltage and current is not reflected is absorbed into the load (or source). See the nice simulation here.

We only need to consider the transmission line properties of the cable when the *wavelength*, λ, of a frequency component in our signal we care about is a significantly small fraction ($\sim 1/10$) of the length of the cable (or the distance between the input and the output of the cable, if it is not straight). For longer wavelengths, the wave has already reached the far end of the cable (and maybe reflected back and forth!) before it has risen in voltage much at the near end. Recall that the velocity of a wave equals λ times f. So in our case, for a 3 m long line, and velocity of c, we need to be careful about these transmission line effects when $\lambda = c/f$ is around or below 30 m, which is when f is above $c/\lambda = 3\text{E}8 \text{ m/s}/30\,m = 10\,\text{MHz}$. Since our signal bandwidth was assumed to be 1.6 GHz, meaning we'll care about frequencies up to 4 GHz or so, there's a large range of frequencies (from 0.01 to 4 GHz) in our example where we need to be careful about transmission line effects.

Skin Effect and Dispersion

Other strange things happen at higher frequencies as well. A wire carrying alternating current has far more current being able to flow near the outer boundary of the conductor than towards the inside. Most of the current occurs within a distance δ from the outside of the conductor, the *skin depth*. More quantitatively, the current density falls exponentially, like $e^{-\delta x}$, as you move a distance x from the outer surface of the conductor. This is called the *skin effect*. The *skin depth* depends on the frequency of the current like $1/\sqrt{f}$. So at high frequencies almost all current flows in the skin of the conductor. This holds even up to frequencies of visible light ($\sim 1 * 10^{14}\,\text{Hz}$) — you know that light doesn't make it too far into a metal — it is reflected! (Above some critical frequency for a material, it does become transparent, because the electrons basically can't keep up. For instance, most metals are transparent to X-rays, and some even in the ultraviolet.) At 1 MHz most of the current flows in the outer $\sim 0.1\,$mm for Al or Cu. At 1 GHz it is $\sim 2\,\mu$m! This is one reason why we'll use thin ground (and power) planes, and thin, wide, copper strips for our "wires" on our PCBs which deal with higher frequencies.

Also, beware that the wires will not carry all frequency components at the same speed, a wave effect called *dispersion*, since the effective L and C of the cable can be a function of frequency. This comes from things like the skin effect and the fact that dielectric materials are usually frequency dependent. It leads to a broadening of the signal as it travels down the wire, since different frequencies will arrive at different times. This also means that the characteristic impedance of the line may be frequency dependent, which is usually a bad thing, since then there are going to reflections no matter how you terminate the cable. And in reality there would also be resistive and conductive losses which decreased the signal size as it traveled down the wire, since the wire has some small resistance per unit length and leakage conductance between the wires per unit length. These losses would also be frequency dependent — in general high frequencies are attenuated more strongly.

Grounding and Decoupling

Another important effect to consider for higher frequency signals, is that the "ground" of our device won't all be at the same voltage anymore! Why? They are all connected together with wire, which we're even assuming can be zero resistance. The problem is that the wire (or conductor of any shape) connecting one ground location to another ground location is going to have inductance. The self-inductance of a wire is proportional to the length of the wire, and also depends a bit on the thickness of it. Imagine two ground points connected by a 5 cm wire, which we'll generously take to be 1 cm in diameter, the self-inductance is ~20 nH. Remember that the impedance of this wire will rise with frequency, $Z = iwL = i*2\pi*f*L$. So for $f = 100\,\text{MHz}$ and $L = 20\,\text{nH}$, $Z = i * 2\pi * 0.1\text{e}9 * 20 * 10^{-9} =$ ~10i Ω. Wires (even perfect ones) have significant impedance at high frequencies!

Let's see the effect of such a ground inductance on the ground voltages. We said that the input signal in our example rose 1 V in 0.2 ns, and the input capacitance was 10 pF. Since $I = C\,dV/dt$, the current from the signal generator was 10 pF $* 1\text{V}/0.2$ ns $= 50$ mA during the rise (and fall) of the signal. This current goes from the signal generator, into the (positive) wire, and into the oscilloscope input. But then where does it go? It has to flow back to where it came from, through the grounding conductors, to make a circuit, roughly following the path of least *impedance*. This current from one point in the grounding conductors to another point in the grounding conductors is called *return current*.

By $V = IZ$, this return current would induce a voltage across this ground wire of 10i $\Omega * 50$ mA $= 0.5$ V (out of phase with the current)! In other words, if we were so careless, the "ground" near one point in our device could be 0.5 V different from the "ground" at another place in our device (temporarily, for 0.2 ns while the current is flowing). Since we usually measure voltage with respect to ground, this could make a 1 V signal look like a 1 V signal in one place and 1.5 V somewhere else! To combat this issue, you must use the shortest ground wires possible (thus make your circuit as small and compact as possible). It's also much better to use a ground *plane*, which has much lower inductance, instead of wires. A ground plane will have just ~1 nH of inductance, at least an order of magnitude better than wires. We'll almost always use a ground plane when designing PCBs (discussed later).

We also have to worry about inductance in our voltage power rails. Imagine a transistor, being fed 1 V at the collector relative to the emitter, and asked to switch on quickly, again in 0.2 ns, so a current of 50 mA. If the 1 V of power is being fed to the transistor through a similar 5 cm wire, this fast current burst would have a similar effect on the voltage of the 1 V "power rail", measured at the transistor. The voltage on the power rail would dip by 0.5 V during switching, so called *switching noise*. This could have deleterious effects on the operation of the transistor and its output, relative to what was expected if the power rail remained at a constant 1 V. And if multiple transistors share the same power rail in the same location (such as in a larger IC), the switching of one transistor can affect the performance of the others. Worse yet, often multiple transistors of an IC will try to switch at the same time (such as on a data bus), leading to a multiplication of the voltage drop.

To maintain the voltage of the power rail during switching, it's critical to ensure that sufficient capacitance is available *locally* between each ground and power (or any other voltage reference) point you care about holding steady, so that the voltage at that point can not change quickly because $dV/dt = I/C$. It's critical that the capacitance be local to avoid the issue of inductance between the capacitance and the point of the circuit whose voltage you are trying to stabilize, since that inductance will lead to exactly the voltage drops discussed above. In practice, these *decoupling* (also called *bypass*) capacitors are placed as close as possible to circuit elements that will have fast rise or fall times, connected between the power and ground pins (or pads) of the component. The amount of capacitance required is determined by the amount of voltage drop that can be tolerated (V), the amount of current required (I), and how long (t) that current is required for: $C > It/V$. For instance, if we want our 50 mA of current for 0.2 ns to cause only a 1 mV voltage drop between power and ground, we would need at least a capacitance of 50 mA $*$ 0.2 ns/1 mV $=$ 10 nF.

Even when placing a capacitor as close as possible to the fast-switching circuit component, the decoupling will be limited by the self-inductance of the capacitor (and that of the power and ground leads on the chip). This shows the total impedance of the decoupling capacitor as a function of frequency for some capacitors:

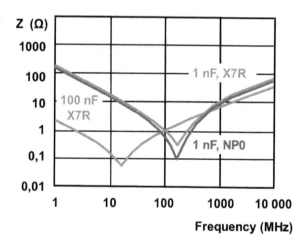

The impedance falls as the frequency increases, since impedance is proportional to $1/f$ for capacitance. But then the impedance starts rising again after the resonant frequency ($1/2\pi\sqrt{LC}$), because the impedance from the self-inductance of the capacitor is proportional to f! Ideally you would have zero impedance between the voltage point you wanted to hold constant and ground, for all frequencies above zero, but it's not possible to do so. Using a bigger capacitor lowers the impedance at lower frequencies, but it tends to be physically larger, which means larger self-inductance and thus larger impedance at high frequencies. Often a multi-capacitor strategy is used, with two or three different types of capacitors being used, with different capacitances and self-inductances, to keep impedance low across a larger range of frequencies. For a realistic device, a whole army of decoupling capacitors, in different places, with a range of values, is needed, forming a "power delivery network (PDN)". "Bulk" capacitors (usually electrolytics, for large capacitance) are placed near the power supply output, and "multilayer ceramic capacitors (MLCC)" are placed near components (for low impedance). Finally, ICs have their own internal capacitance, at even lower inductance, but the capacitance is small.

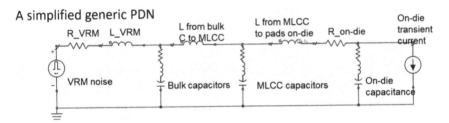

Oscilloscopes and Probes

The issues discussed in this chapter have some important consequences for how we *measure* higher frequency signals, or fast signals with small rise or fall times. If we're not careful about how we use the oscilloscope, we won't be accurately measuring our signal — we'll be looking at artifacts of the oscilloscope and its probe!

The most common type of probe is the 10× *probe*. Often these come as 1 × /10× probes, with a little switch on the probe to change between 1× and 10× mode. Let's study the 10× mode first, since the 10× *mode is the mode you should usually use*. A *simplified* diagram of the probe, cable, and scope input is shown below. The output impedance of the source is assumed here to be 50 Ω, but of course we don't know what the source impedance is going to be, since it depends on the device you're measuring. The voltage at point A is being measured and is input to the scope at point B. The scope has an input *resistance* (load) of 1 MΩ (typically) and *capacitance* of 20 pF (varies from 10 to 25 pF depending on the scope). Some scopes (good ones) will also be able to switch in 50 Ω input impedance as well, needed for matching the characteristic impedance of 50 Ω cable, but when using the 10× probe, you must be in 1 MΩ mode. The probe (in 10× mode!) has a series 9 MΩ resistance, in parallel with an adjustable capacitance, C_{comp}. There's also the capacitance of the probe cable, which is an additional load of about 100 pF (typically). At DC, this is just a simple voltage divider, so $V_B = V_A * R_{in}/(R_{div}+R_{in}) = V_A/10$. *Note that* 10× *probe mode actually divides the signal voltage by* 10, so they really should call it the "10/" probe, but oh well. At high frequency, the impedance of the capacitors will be much smaller than that of the resistors, so we can just consider them. For instance, the 20 pF capacitance, at 1 MHz, is just $1/2ifC = {\sim}10\,\text{k}\Omega$. So at high frequency, the circuit is again just a voltage divider, but using *reactance* instead of resistance. As long as the reactance of C_{comp} is also 9 times that of $C_{cable} + C_{in}$, the V_B will still be $V_A/10$ at high frequency, and in fact at *any* frequency. This trick is generally called *compensation*. In order for the reactance to be the right ratio, this means that the capacitance of C_{comp} must be 1/9 that of $C_{cable} + C_{in}$, since reactance is inversely proportional to capacitance, hence $C_{comp} = (100\,\text{pF} + 20\,\text{pF})/9 = {\sim}13.3\,\text{pF}$. It's left adjustable because there are going to be other stray capacitances at the ∼1 pF level, which we'll need to adjust C_{comp} for, in situ.

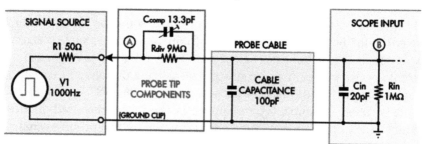

Here's how to adjust, or *compensate*, a 10× probe for an oscilloscope. The oscilloscope usually has a 1 kHz square wave output on the front panel near the BNC inputs. Attach the probe tip (using the springy clip) to this, and the ground clip of the probe to the ground terminal just underneath the square wave output. Recall that this square wave will have low frequency components (1 kHz and below), but also high frequency components from the fast rise and fall times of the square wave. Also keep in mind where the crossover frequency is for the voltage divider, from resistive to reactive. It will be about where the reactance of C_{comp} equals the resistance of R_{div}, which is when $1/2\pi f C_{\text{comp}} = 9\,\text{M}\Omega$, so f = ~1.4 kHz. If C_{comp} is not just right, the high frequency components of the square wave will be either too large or too small, but if it's just right, it'll look like a nice square wave, as shown below. So just adjust the screw on the adjustable capacitor (with the supplied *plastic* low-capacitance screwdriver!) until it looks nice and square.

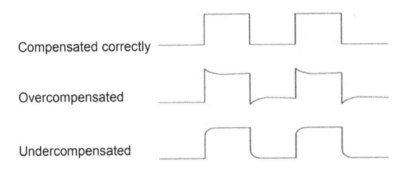

Compensated correctly

Overcompensated

Undercompensated

What is the *load* from the oscilloscope and probe on the circuit under test? Ideally, the input impedance would be very large, for all frequencies. Otherwise the measurement device will have a large effect on the circuit being probed, which is bad. For our 10× probe, the load impedance at DC is 10 MΩ — very good. But beware that because of the capacitance to ground (~120 pF), the input *impedance will decrease* linearly with frequency once we're above that ~1.4 kHz where we are relying on a reactive voltage divider. By 1 MHz, the impedance is already down to $1/2\pi f C = 1/(2\pi 1\,\text{MHz}\,120\,\text{pF}) =$ ~1.4 kΩ! The bandwidth of our 10× probe should be very good though, limited only by the low pass filter formed from the ~50 Ω source impedance and the ~10 pF of capacitance of the probe tip, which has a −3 dB frequency of ~300 MHz.

However, it's not quite so simple to achieve this 300 MHz bandwidth. At high frequencies, we also need to consider the transmission line effects of the probe cable. These will become important for frequencies above 30 MHz or so, for a 1 m cable (recall similar calculation above). As we saw, to prevent reflections the load termination impedance should match the characteristic impedance of the transmission line cable, $Z_0 = \sqrt{L/C}$. But that's not what we have at all. The cable has an impedance of $\sim 50\,\Omega$, and our termination is $1\,\mathrm{M}\Omega$ at DC and decreases with frequency due to the input capacitance ($\sim 120\,\mathrm{pF}$). The solution, devised in 1969 at Tektronix by Joe Weber, is to use a *lossy* transmission line, with about $\sim 100\,\Omega$ of resistance in the cable. If you measure the resistance of your $10\times$ probe from one end to the other (of the signal wire in the middle of the cable), you'll find about $100\,\Omega$, not close to $0\,\Omega$! This effectively absorbs the reflections in a steady way without causing any other impedance mismatch points which would cause other reflections themselves. (Recall that at $\sim 30\,\mathrm{MHz}$, the input impedance of the system is only $\sim 50\,\Omega$, so $100\,\Omega$ of series impedance will have a significant damping effect.) In addition, it helps to add a bit of terminating load (R_{comp}) at the scope input. Of course, we can't have that load there at low frequencies since it would mess up our voltage divider, but we only need it at high frequencies, when the transmission line effects are important. If we put R_{comp} in series with a small capacitor, it will only have an effect at high frequencies when that capacitor has small impedance. It's chosen to be more than $50\,\Omega$ ($\sim 68\,\Omega$) because the characteristic impedance of the cable is effectively larger at these high frequencies due to the resistive lossiness of the cable we've included. The full picture of the source, probe, and scope input is shown below. A "normal" $10\times$ probe will have a $-3\,\mathrm{dB}$ bandwidth of $\sim 100\,\mathrm{MHz}$, and good ones (that are tuned quite carefully and minimize capacitance etc.) can go up to $\sim 500\,\mathrm{MHz}$.

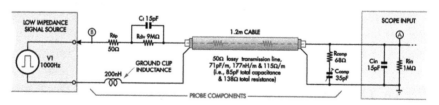

A critically important thing to be aware of in the $10\times$ probe is the "ground clip". This is typically a $\sim 5\,\mathrm{cm}$ long wire that you can use to connect the "ground ring" of the probe to a "ground" point on your device being measured. This wire, plus the part of the tip before getting to the cable, can typically have $\sim 200\,\mathrm{nH}$ of inductance. This forms a RLC resonant circuit with a frequency of $1/[2\pi * \sqrt{(L * C)}]$, which for C of $10\,\mathrm{pF}$ of tip capacitance is at ~100 MHz. With a resistance of $50\,\Omega$ for the source, this gives a quality factor of $Q = 1/R * \sqrt{(L/C)} = \sim3$, so quite a bit of ringing at that frequency. Recall that inductance is proportional to length (and loop inductance is proportional to area), so using a shorter wire that also makes a smaller loop (from probe ground to board ground to signal source to probe tip) will help a lot, and the resonant frequency will get pushed beyond the bandwidth of the probe and scope, leading to less ringing, as shown in the following figure.

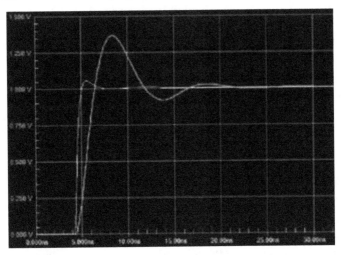

Also, note that you can't attach the "ground" wire of the scope probe to *any* location in your circuit, if your circuit is "plugged into the wall", meaning the ground of your circuit is attached with low resistance to the ground of the electrical wiring in the wall ("earth"). Your oscilloscope is also plugged into the wall and connected to earth ground, so the ground lead on your oscilloscope is connected to earth ground. If you put the ground lead of your oscilloscope on some point in your circuit that is not "ground" (at the same voltage as the earth ground in the wall), a current (possibly very large!) will flow from your circuit through the ground of your oscilloscope probe and back into the wall and then back to your circuit. You have shorted your circuit to ground and possibly damaged your circuit and/or probe and/or scope in the process! Your scope is *not* a multimeter, which is powered by batteries and thus not connected to earth ground — a "floating ground". If you really need to measure the difference in voltage (vs time) using your scope between two non-ground points in your circuit, you need to get a "differential probe" which isolates your probe ground from the oscilloscope ground, or a battery-powered oscilloscope (they are made, but usually low performance), or power your oscilloscope from an isolating transformer or inverter.

So what about "1×" mode on the probe? All this does is short out across C_{comp} and R_{div}, so it is no longer a voltage divider (and thus no compensation is needed). It is now basically just a length of *lossy* ~200 Ω cable with ~100 pF of capacitance (to ground) going into the scope input with 1 MΩ load and ~15 pF capacitance (to ground). Since there is no voltage divider, 1 V on the input will give you 1 V at the input to the scope. This is the one benefit to 1× mode — the signal is not diminished in amplitude (allowing you to see smaller signals on the scope). But what about the bandwidth? The RC low-pass filter −3 dB frequency of this ~200 Ω resistance in the cable with the ~120 pF in the cable and scope input is about 6 MHz! It is useless for high frequency signals! (On the plus side, we don't have to worry about reflections!) It is worth repeating — BEWARE — the bandwidth of the probe in 1× mode is low — don't use it unless you have to, to see a small signal!

What if you want more bandwidth than the 10× probe offers? The cheap (but pretty good!) option is the "950/50 Ohm probe", which is easily made by soldering a 950 Ω (or 1 kΩ) resistor onto the signal (center) wire of the cut end of a 50 Ω coax cable. Then the ground shield of the cable is attached (with as short a wire as possible!) to the ground of the circuit. Then just be sure to terminate the cable with 50 Ω load at the input to the scope, to eliminate reflections. If your oscilloscope does not have a 50 Ω impedance option, you can use a BNC adapter that goes on the end of the cable at the scope that includes the 50 Ω load, or solder a $1/4$ W 50 Ω (or 47 Ω, since 50 is hard to find!) resistor between the center wire and ground shield of the cable very near the BNC connector. This forms a 20:1 (or 21:1) voltage divider between the resistor and the impedance of the cable. You can use a 450 Ω resistor if you want to keep the 10× probe signal amplitude reduction. The bandwidth is easily a ~few GHz, as long as you're careful about parasitic capacitance of the 1 kΩ resistor (find one as small as possible!) and keep the resistor a bit away from other conductors (especially ground planes) to reduce stray capacitance.

At $50\,\Omega$ input impedance, $1\,$pF of capacitance will limit your $-3\,$dB bandwidth to $3\,$GHz.

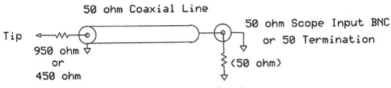

50 ohm Coaxial Line

Tip — 950 ohm or 450 ohm

50 ohm Scope Input BNC or 50 Termination

(50 ohm)

Lo-Z Resistive Passive Probe

The downside of the inexpensive $50\,\Omega$ passive probe is that the signal amplitude is reduced by a factor of 10 or 20, the bandwidth is limited by stray and parasitic capacitances, and the load on the circuit under test is large, $\sim 1\,$kΩ. For higher bandwidth, larger signal sensitivity, and lower load, you need to get an *active probe* which (typically) uses a FET amplifier in the tip of the probe. These are expensive, like $3000 (used!) for a $3\,$GHz, $1\,$MΩ, $0.8\,$pF, probe. Hopefully by now you have some appreciation for why it's hard to design and manufacture such a thing, with flat frequency response out to 3 GHz, where even "half a puff" ($0.5\,$pF) of stray capacitance would ruin the whole thing.

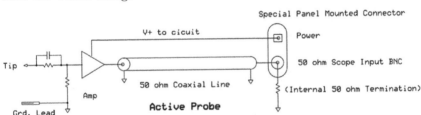

Special Panel Mounted Connector

V+ to cicuit

Power

Tip

Amp

50 ohm Coaxial Line

50 ohm Scope Input BNC

(Internal 50 ohm Termination)

Grd. Lead

Active Probe

Your oscilloscope will have a bandwidth rating. This is usually driven by two things: (1) the analog bandwidth of the input channel "front-end" electronics just behind the BNC input and (2) the sample rate of the ADC that is digitizing the analog signal. We've already discussed various things that can limit analog bandwidth. But we'll just mention here that the digitization rate will also effectively limit the bandwidth. Of course we can not see a rise time of 0.2 ns if we are only sampling the signal every 1 ns (at 1 G sample-per-second, or 1 Gsps). There's a theorem called the Nyquist theorem that states that the best you can do is see a frequency that's $1/2$ of your sampling frequency. So sampling at 1 Gsps we may be able to "see" a sine wave of 500 MHz. That would give one sample at the peak of the wave and one at the bottom, so you could only reconstruct it as a sine wave if you knew to begin with it was a sine wave — not so useful. Higher frequencies, by the way, would be *aliased*, meaning you would see some fraction of the wave, and interpret it as a lower frequency wave. For instance, a 1 GHz wave would appear (with our 1 Gsps sampling rate) as a 0 MHz wave — we would see the same voltage for each sample. In general, an aliased signal will appear at |sampling freq – signal freq|, so 600 MHz would appear as 400 MHz! The same effect occurs for wheels spinning faster than the 30 Hz frame rate a move is shot at — it can appear to be rotating slowly, or even backwards. missing every other oscillation. Oscilloscopes will usually make sure that their analog bandwidth will kill off any frequencies above the Nyquist frequency, or add an *anti-aliasing* low-pass filter to remove those higher frequencies. In practice, you usually want the sampling rate of the oscilloscope to be at least ~10 times the stated analog bandwidth (listed on the front panel of the scope). So, a 100 MHz scope should have a sampling rate of 1 Gsps. Beware that many scopes will "interleave" the sampling between the channels, so if you are using a 4 channel scope but have only one channel active, you may get the full 1 Gsps, but if you turn on a second channel they will each get 500 Msps, and for four channels active you will get 250 Msps — beware!

Cables

As we've seen, to transmit signals over distances greater than ~1/10th of the smallest wavelength of the signal we care about, we need to use a transmission line. And we need to be careful not to pick up noise from external fields along the way. There are at least two main types of "noise-resistant" transmission line cables in common use.

The first is *coaxial cable*, consisting of a central conducting wire and an outer "ground" shield, with a dielectric material (non-conductive) in between. These can be made with various impedances from ~25 to ~150 Ω (depending on the dimensions and the dielectric), but the most common is 50 Ω (and then maybe 75 Ω). It has large bandwidth, up to 10 GHz or more, often limited by the connectors, which must maintain the line impedance. The smaller SMA connectors have higher bandwidth ($\sim$15 GHz) than the bigger BNC ($\sim$few GHz). You'll find these cables whenever you want the highest bandwidth and lowest signal loss and dispersion possible, and where cost and space are not an issue.

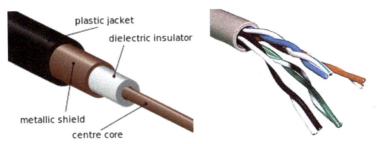

The other category is twisted-pair cables, which are (as the name would imply) pairs of wires twisted together. Often several pairs are grouped within a single cable, e.g. Cat5 cable has four pairs of wires. The number of twists per unit length and the thicknesses and spacings of the wires can be controlled pretty well, to give impedance of $\sim$50 to $\sim$300 Ω, but the most common is 100 Ω. It is much cheaper and more flexible than coaxial cable, but the bandwidth is smaller, less than $\sim$1 GHz usually. Often there is no (grounded) shield conductor around the cable, but noise from external fields is reduced in several ways. One wire of each pair is usually a ground wire and thus effectively shields radiation of wavelength greater than the length of one twist. Also, the loop area of each pair is small because the wires are kept close together, so inductance is minimized. Lastly, any lower frequency interference is experienced equally by both wires of the pair, so does not affect the *difference* in voltage between the pairs. Differential signaling is used for Ethernet and LVDS, for example (discussed later).

Lastly, for digital signals, another common solution is fiber-optic cable. The frequency of (red) light is around 400 THz, so there are no bandwidth limits from the cable itself. There are limits to how fast you can pulse a laser diode on and off, which must be driven by electrical inputs and thus subject to the stray capacitance limits we discussed above. But they can easily send 10 Gbps or more today, per fiber, per color. They have the advantage of allowing no DC electrical connection (eliminating ground loops and other problems), as well as having very low signal loss over long distances. The only downsides are cost and durability.

Lab: High-speed Signals

(1) Send $1\,V_{pp}$ pulses (as short in duration as possible, with as fast a risetime and falltime as possible) from the pulse generator into the oscilloscope, using a $50\,\Omega$ coax cable.

   Study the shape of the pulse with various low-pass filter cut-off frequencies applied. (The filters can be created in the oscilloscope's channel menu.) How does the risetime (or falltime) depend on the bandwidth of the filter?

   Are there over- and under-shoot artifacts on the edges of the pulse waveform? How do they depend on the filter bandwidth?

(2) Now use a scope probe in $10\times$ mode to measure the signal, on the breadboard, instead of a direct coax cable attached to the scope. How is the bandwidth of the signal impacted? The rise and fall times of the signal edges?

   How do these change when using $1\times$ mode for the probe?

   What if you use a bad (i.e. long) scope probe ground connection (or none at all, other than through the power plug)?

(3) Now send the pulses from the signal generator down a long coax cable ($\sim$40 + feet) and measure the signal at both the near end and the far end, using $10\times$ scope probes on both ends. Do this with an open far end (high impedance), shorted far end, and properly terminated far end. (You can use a coax cable tee at the near end, and the breadboard BNC connector at the far end.)

   Most signal generators are nice and have $50\,\Omega$ output impedance. What if you break this impedance matching by adding a $1\,k\Omega$ resistor in series with the signal generator output right near the output?

(4) Attach a coax cable tee to the signal generator output to send the pulses down two long cables and observe the signal at both the near end and one of the far ends (using $10\times$ scope probes), using proper termination at both of the far ends. (You can use a second coax tee at the near end, and the breadboard BNC connectors at the far ends.) What happens without proper termination, either open or short, at one of the far ends?

Chapter 7

Microprocessors

Often you want to accomplish some fairly simple task like: monitor some temperature, if it's too warm, turn on some cooling fan, and if it remains too warm for too long or passes some threshold, send a warning that there's some problem with the temperature. This sort of thing could be done with an analog circuit (how?), or a mix of analog and discrete digital logic chips (how?), but these days you would use a *microcontroller*.

A microcontroller is a very small "computer" with a CPU that runs a program, usually just a single program that you have programmed it with. It starts this program when it is powered on, and it keeps running it until it is powered down. What makes them very useful is that they cost only ~\$1, fit on a single chip, use just ~25 mA (at ~5 V) of power, and come with many ways of interfacing with the real world. They have:

- digital outputs: metal pins that can be set to output a "high" voltage (equal to the power supply voltage, ~5 V) or "low" voltage (GND);
- digital inputs: metal pins that when driven by a high or low voltage will register a 1 or 0 to the program;

- analog inputs: metal pins that will send a value between 0 to 1023 (for a 10 bit precision) to the program depending on the input voltage at a pin, using a built-in analog to digital converter (ADC);
- analog outputs: metal pins that will output a voltage (using pulse-width-modulation (PWM) usually!) depending on a value between 0 and 1023 (for 10 bit precision) sent to it by the program;
- ability to talk to other digital devices using "serial interfaces" such as UART, I2C, or SPI.

Of course microcontrollers also have severe limitations compared to standard computers, such as:

- slow speed: they run at just ~16 MHz or so (compared to ~4000 MHz for your laptop), but that's still plenty fast enough to do quite a bit in a single millisecond (~16,000 clock cycles);
- small storage: they have just ~32 kB or so of flash memory (compared to ~256,000,000 kB for your laptop), which has to store your entire program (and the ~2 kB "boot loader" that loads your program at startup);
- little "RAM": they have just ~2 kB or so (compared to ~8,000,000 kB for your laptop), which has to hold the program stack and all the variables your program uses at runtime.

Usually you don't use a microprocessor directly (especially at the beginner level). You use a *development board,* which breaks out the pins from the actual microprocessor chip to easily connectable header pins. For instance, see the diagram below of a classic microprocessor board, the Arduino Uno, which uses the ATMEGA328P microprocessor. It provides regulated power to the chip from an external supply, adds an oscillator to provide a clock signal to the chip, has extra interface circuitry for a USB connection, and has a reset button and some LEDs.

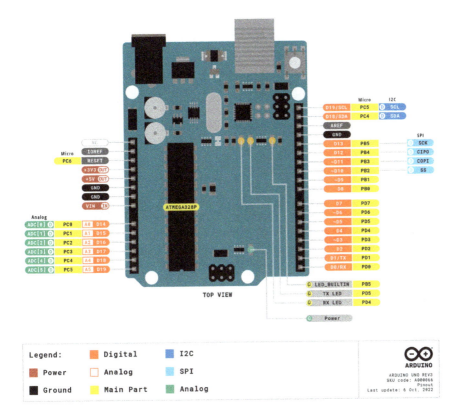

TOP VIEW

Legend:		Digital		I2C
	Power		Analog		SPI
	Ground		Main Part		Analog

ARDUINO
ARDUINO UNO REV3
SKU code: A000066
Pinout
Last update: 6 Oct. 2022

There's thousands of microprocessors available, but we'll focus on those compatible with the "Arduino" software tools, which use a C-like programming language and IDE (integrated development environment). These are easy to program and can also make use of a huge set of available 3rd-party libraries. The classic is the Arduino Uno, shown below. It has 14 digital input/output pins, of which 6 can be used as analog (PWM) outputs, 6 analog inputs, a 16 MHz ceramic resonator, a USB connection, a power jack, an LED, and a reset button.

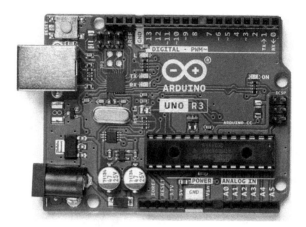

First you'll need to install the Arduino software on your computer. From that software you can write programs, compile them, and upload them to the Arduino board. Upon starting the Arduino software, you see a text editor with some options, as shown in the following figure.

Here you write your "sketch", which is what an Arduino program is called. Once you plug the Arduino into your computer with the USB cable, you should make sure that "Arduino Uno" is selected under Tools -> Board and that the right port (like COM5) is selected under Tools -> Port. Then you can simply compile and upload the sketch to the board using the Sketch -> Upload command, or the -> button. The simplest sketch is the "Blink" sketch. It just blinks the built-in LED once per second. You can get it from the File -> Examples -> Basics -> Blink menu. The code is also shown below.

```
// the setup function runs once when you press reset or power the board
void setup() {
  // initialize digital pin LED_BUILTIN as an output.
  pinMode(LED_BUILTIN, OUTPUT);
}
// the loop function runs over and over again forever
void loop() {
  digitalWrite(LED_BUILTIN, HIGH);  // turn the LED on
  delay(1000);                      // wait for a second
  digitalWrite(LED_BUILTIN, LOW);   // turn the LED off
  delay(1000);                      // wait for a second
}
```

This sketch has two functions that every Arduino sketch also must have. The first, "setup", is run once at the start of the program, when the Arduino is powered on. The second, "loop", then runs again and again until the power is turned off. An Arduino program never ends! For this Blink program, the setup function just does one thing — sets the LED_BUILTIN pin (which is pin 13 on the Arduino Uno) to be a digital output. Each digital pin can be set as a digital output or digital input.

The loop function here uses the digitalWrite function to set the state of a digital output pin. It will set the LED_BUILTIN pin (13) to be high (5 V) or low (0 V). This will turn the connected builtin LED on or off, respectively. Beware output current limitations — it depends on the microprocessor and board, but generally keep it to less than $\sim$20 mA per output. Here there is a $\sim$1 k resistor in series with LED so the current will be less than $I = V/R = 5\,\mathrm{V}/1\,\mathrm{k} = 5\,\mathrm{mA}$. The "delay" function is used to wait (approximately!) a given number of milliseconds between lines of code. Once the function finishes, it is immediately called again, so the LED will blink forever, until the board is powered off.

Let's look at another example that shows how to read a digital input pin, and print output to the computer screen. It is at

File -> Examples -> Basics -> DigitalReadSerial, and shown below. In setup, we set pin 2 as an input. Then we initialize the Serial output stream, at a *baud rate* of 9600 bits per second. In the loop function, we read the state of the digital input pin. If the voltage on the pin is high (near 5 V), buttonState will equal 1, but if it's low (near 0 V), buttonState will equal 0. We then output the value of buttonState to the Serial output stream, which will let us see the output on our computer. To see the output go to Tools -> Serial Monitor, and make sure to set the baud rate at the bottom of the window to 9600 baud, to match the setting in the sketch.

```
void setup() {
  // make the pushbutton's pin an input
  pinMode(2, INPUT);
  // initialize serial communication at 9600 bits per second
  Serial.begin(9600);
}
void loop() {
  // read the input pin
  int buttonState = digitalRead(pushButton);
  // print out the state of the button
  Serial.println(buttonState);
  delay(1);  // delay in between reads for stability
}
```

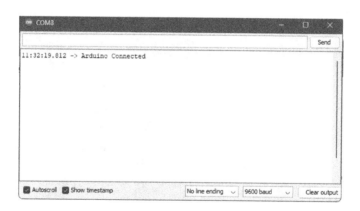

There's one issue with digital inputs though — if nothing is connected to it, like when the button is open, what voltage is the input pin? It is *floating*. Is the input a 0 or a 1? It's nice to have a default state. Calling pinMode(2,INPUT_PULLUP) solves this by connecting a ~10 k input pullup resistor (inside the microprocessor!) from the input to the VCC (5 V) rail, as shown in the following figure. Then if nothing is connected, the input will be pulled up to 5 V, and will register as a 1 in the code.

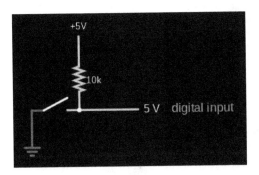

We can easily read an analog voltage as well, as shown in the File -> Examples -> Basics -> AnalogReadSerial example, shown below. The microprocessor contains a builtin analog to digital converter (ADC), which takes an (analog) voltage between 0 and 5 V and outputs a (digital) number between 0 and 1023, linearly proportional to the input voltage. This is a 10-bit ADC which means there are $2^{10} = 1024$ possible output values. For example, a 1 V input would give about $1024 * 1\,\text{V}/5\,\text{V} = \sim205$. It takes about 0.1 ms to read the voltage.

```
void setup() {
  // initialize serial communication at 9600 bits per second
  Serial.begin(9600);
}
void loop() {
  // read the input on analog pin 0
  int sensorValue = analogRead(A0);
  // print out the value you read
  Serial.println(sensorValue);
  delay(1);  // delay in between reads for stability
}
```

It's also *sort of* possible to output an analog voltage (between 0 and 5 V) to a pin. On the Arduino Uno, pins 3, 5, 6, 9, 10, and 11 are capable of this, using the analogWrite function, as shown below. There are 8 bits of precision, so values from 0 to 255 can be input, which are linearly mapped to output voltages of 0–5 V.

```
void setup() {
  // nothing happens in setup
}
void loop() {
  // fade in from min to max in increments of 5 points
  for (int fadeValue = 0; fadeValue <= 255; fadeValue += 5) {
    // sets the value (range from 0 to 255):
    analogWrite(9, fadeValue);
    // wait for 30 milliseconds to see the dimming effect
    delay(30);
  }
}
```

The reason it is only "sort of" an analog output is that the microprocessor is actually only outputting 0 V or 5 V at any given moment, not the requested intermediate voltage! However, if a value of 1 V is requested, the output will be on (at 5 V) for about 1/5 of the time and off (at 0 V) the rest of the time. This means that the average output voltage is about 1 V. This method of adjusting the *duty cycle* of an output in order to control the mean output voltage is called pulse width modulation (PWM), as shown below. On the Arduino Uno, the frequency of the modulation cycle is 490 Hz by default on most pins, and 980 Hz on pins 5 and 6.

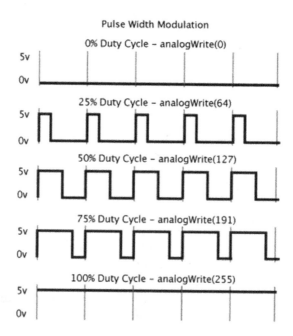

In many cases, it is necessary to smooth this output into a steady DC voltage. Fortunately, by now we know how to do this — use a low pass RC filter, as shown in the following figure. In the following example, the cutoff frequency is $1/2\pi RC = 1/2\pi * 1 \text{ k}\Omega * 10 \text{ uF} = \sim16\,\text{Hz}$. This is well below the 490 Hz modulation frequency, so the output is pretty smooth. The larger R or C, the lower the cutoff frequency. Using a larger R will increase the output impedance though, making it more susceptible to effects from the input impedance of the following stage. (Remember that the input impedance of the following stage, whatever the analog output is attached to, should be 10× the output impedance!) So using a larger capacitor, instead of a larger resistor, might be needed. Or you could use an op-amp buffer after the analog output to provide a large input impedance, and a low output impedance to the following stage.

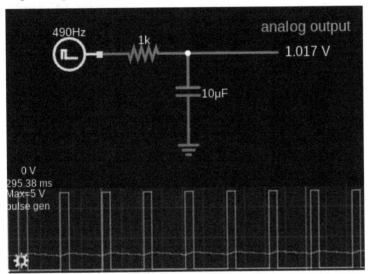

One more thing we often want to do is control the Arduino using some commands from the computer. For example, this code below, from File -> Examples -> Communication -> Dimmer, reads a byte (8 bits) from the serial *input* and then sets the brightness of the LED accordingly.

```
void setup() {
    // initialize the serial communication
    Serial.begin(9600);
    // initialize the ledPin as an output
```

```
  pinMode(LED_BUILTIN, OUTPUT);
}
void loop() {
  byte brightness;

  // check if data has been sent from the computer
  if (Serial.available()) {
    // read the most recent byte (which will be from 0 to 255)
    brightness = Serial.read();
    // set the brightness of the LED
    analogWrite(LED_BUILTIN, brightness);
  }
}
```

The byte to send over serial can be entered using the Tools -> Serial Monitor dialog, typing a character into the input box, and pressing send. This might not behave quite as you expect though. For instance, entering a "0" and sending will not set the LED brightness to 0. And sending "255" actually sends 3 bytes, not just 1 with a value of 255. When you send "0", what this really does is send the ASCII value of 0, which is 48 in decimal, according to the ASCII table, shown as follows.

Dec Hx Oct Char		Dec Hx Oct Html Chr	Dec Hx Oct Html Chr	Dec Hx Oct Html Chr
0 0 000 NUL (null)		32 20 040 Space	64 40 100 @ @	96 60 140 ` `
1 1 001 SOH (start of heading)		33 21 041 ! !	65 41 101 A A	97 61 141 a a
2 2 002 STX (start of text)		34 22 042 " "	66 42 102 B B	98 62 142 b b
3 3 003 ETX (end of text)		35 23 043 # #	67 43 103 C C	99 63 143 c c
4 4 004 EOT (end of transmission)		36 24 044 $ $	68 44 104 D D	100 64 144 d d
5 5 005 ENQ (enquiry)		37 25 045 % %	69 45 105 E E	101 65 145 e e
6 6 006 ACK (acknowledge)		38 26 046 & &	70 46 106 F F	102 66 146 f f
7 7 007 BEL (bell)		39 27 047 ' '	71 47 107 G G	103 67 147 g g
8 8 010 BS (backspace)		40 28 050 ((	72 48 110 H H	104 68 150 h h
9 9 011 TAB (horizontal tab)		41 29 051))	73 49 111 I I	105 69 151 i i
10 A 012 LF (NL line feed, new line)		42 2A 052 * *	74 4A 112 J J	106 6A 152 j j
11 B 013 VT (vertical tab)		43 2B 053 + +	75 4B 113 K K	107 6B 153 k k
12 C 014 FF (NP form feed, new page)		44 2C 054 , ,	76 4C 114 L L	108 6C 154 l l
13 D 015 CR (carriage return)		45 2D 055 - -	77 4D 115 M M	109 6D 155 m m
14 E 016 SO (shift out)		46 2E 056 . .	78 4E 116 N N	110 6E 156 n n
15 F 017 SI (shift in)		47 2F 057 / /	79 4F 117 O O	111 6F 157 o o
16 10 020 DLE (data link escape)		48 30 060 0 0	80 50 120 P P	112 70 160 p p
17 11 021 DC1 (device control 1)		49 31 061 1 1	81 51 121 Q Q	113 71 161 q q
18 12 022 DC2 (device control 2)		50 32 062 2 2	82 52 122 R R	114 72 162 r r
19 13 023 DC3 (device control 3)		51 33 063 3 3	83 53 123 S S	115 73 163 s s
20 14 024 DC4 (device control 4)		52 34 064 4 4	84 54 124 T T	116 74 164 t t
21 15 025 NAK (negative acknowledge)		53 35 065 5 5	85 55 125 U U	117 75 165 u u
22 16 026 SYN (synchronous idle)		54 36 066 6 6	86 56 126 V V	118 76 166 v v
23 17 027 ETB (end of trans. block)		55 37 067 7 7	87 57 127 W W	119 77 167 w w
24 18 030 CAN (cancel)		56 38 070 8 8	88 58 130 X X	120 78 170 x x
25 19 031 EM (end of medium)		57 39 071 9 9	89 59 131 Y Y	121 79 171 y y
26 1A 032 SUB (substitute)		58 3A 072 : :	90 5A 132 Z Z	122 7A 172 z z
27 1B 033 ESC (escape)		59 3B 073 ; ;	91 5B 133 [[	123 7B 173 { {
28 1C 034 FS (file separator)		60 3C 074 < <	92 5C 134 \ \	124 7C 174 |	
29 1D 035 GS (group separator)		61 3D 075 = =	93 5D 135]]	125 7D 175 } }
30 1E 036 RS (record separator)		62 3E 076 > >	94 5E 136 ^ ^	126 7E 176 ~ ~
31 1F 037 US (unit separator)		63 3F 077 ? ?	95 5F 137 _ _	127 7F 177 &#127; DEL

Source: www.LookupTables.com

As another example, sending the character "y" would result in a brightness of 121. If you actually wanted to parse an ASCII string, like "149" into the integer 149 to set the brightness of the LED to 149, edit the code to this instead:

```
brightness = Serial.parseInt();
```

The parseInt function will read several ASCII numeric characters and return an integer (after a non-numeric character is read, or a timeout).

How is this serial data actually traveling over the USB cable? That's a bit complicated and has to do with USB, but fortunately there's a chip on the Arduino board that translates between this nasty USB protocol and the very simple *UART* serial protocol. Pins 0 and 1 on the Arduino are the receive (RX) and send (TX) pins, respectively, for UART serial data to and from the computer. These are digital pins, so they are either high or low voltage. The rate that the pins toggle from low to high or high to low is the baud rate. For example, at 9600 baud, each bit cycle would be 1/9600 seconds long, or about 0.1 ms. An example of sending a byte is shown below. First there is always the start bit, which pulls the TX pin low for one cycle. Then the next 8 cycles are the actual bits it's sending, where a 1 would be high voltage and a 0 low voltage. Then there's always a stop bit, which is high voltage for 1 cycle. So to send a byte, it takes 10 bit cycles, or a little over 1 ms at 9600 baud, because of the 2 start and stop bits. (There are variations on this protocol, for instance sometimes 2 stop bits are used, or a parity bit is added, but what we described is most common.)

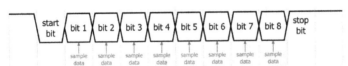

Note that it's critical that the sender and receiver agree ahead of time on the baud rate! And the sender and receiver need to have "clocks" running at approximately the same speed, so 9600 baud means the same actual bit rate on both ends, otherwise the receiver won't be measuring the voltage at the correct times, when the voltage is stable between bit transitions. Receivers are usually flexible enough to handle a ~few % difference in baud rate or clock speed. Going at higher baud rates, like 1 Mbaud, presents other difficulties, like assuring low enough capacitance so that the voltage can change from low to high (or vice versa) quickly enough. The highest common baud rate used is 115,200.

The default "Serial" object on Arduinos uses pins 0 and 1, as mentioned above. You can have additional serial connections over

other pins using the SoftwareSerial *library*, as shown below. There are
tons of libraries available for doing all sorts of things with Arduinos.

```
#include <SoftwareSerial.h>
SoftwareSerial mySerial(2, 3); // RX, TX
void setup()
{
  // set the data rate for the SoftwareSerial port
  mySerial.begin(38400);
}
void loop() // run over and over
{
  mySerial.println("Hello?");
}
```

UART was great for when you have just two things communicating. But what if you want the Arduino to talk to 29 different sensors and gather data from each of them? Having 29 separate UART connections would be unwieldy, requiring 58 wires, and far more input/output pins than the Arduino has. A better protocol is "I2C", which allows one or more "master" devices (like Arduinos) to talk to up to 127 "slave" devices, using just two wires, as shown below. One wire is for "data" (SDA) and the other is for "clock" (SCL). Both of these wires must be pulled up to 5 V using pullup resistors (using physical external resistors, not internal ones), about ~10 kΩ usually, depending on the speed desired and the capacitance of the wires.

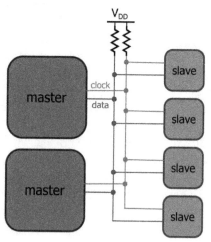

Each transmission has several pieces, as shown below. After the start bit, an address is sent, which selects which slave is being talked to. Each slave on the bus must have a unique address. If and only if its address matches that which was sent, it will react to the rest of the message. Messages can initiate either reads from or writes to the slave. Data frame 1 usually specifies what type of information is being written or read to/from the slave. Data frame 2 usually specifies the value of that data written or read. Acknowledgement bits are sent by the master and slave to make sure they received the data. And finally there's a stop bit.

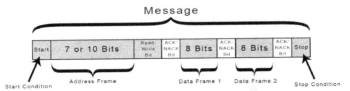

From the Arduino, acting as a master, code such as this below would allow you to write data to a particular slave with address 44, which is fixed by the device. Consult the datasheet of the device you're trying to interface with to see how to set its i2c address. Many have a few pins which can be either low, high, or floating, to assign a different address to each device on the bus. It uses the "Wire" library. Note that the SDA and SCL pins on the Arduino Uno are D18 and D19 (top right of board), respectively, which are the same as A4 and A5, respectively.

```
#include <Wire.h>
void setup()
{
  Wire.begin(); // join i2c bus, using A4 and A5
}
void loop()
{
  Wire.beginTransmission(44); // transmit to device #44 (0x2c)
  Wire.write(byte(0x00));     // sends instruction byte
  Wire.write(17);             // sends potentiometer value byte
  Wire.endTransmission();     // stop transmitting
}
```

Another very popular interface is the Serial Peripheral Interface (SPI). The main advantage is it's faster, but it needs more wires. There is still a clock signal like for i2c, usually called SCLK. Then there are *two* wires for sending data, as opposed to the one for i2c. This allows data to be sent bi-directionally at the same time, called duplex communication. For i2c the master was either sending to the slave or the master was listening to the slave, not both at the same time. The two wires are often called Master Out Slave In (MOSI) and Master In Slave Out (MISO). Lastly there is a separate wire from the master to each slave which allows the master to select that slave (or combination of slaves) for listening to and/or sending to. This is usually called Slave Select or Chip Select (SS/CS). So, it could look like below for interfacing a master with three chips over SPI.

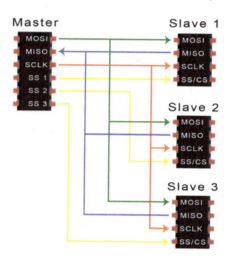

The Arduino has a built in SPI interface, on the pins shown below. Note that some of the pins are duplicated on the "ICSP" header at the bottom — you can use either.

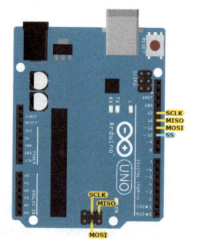

You would then use the SPI library (or another library for a specific device, which includes the SPI library) to talk to a connected device like this:

```
#include <SPI.h>
void setup() {
   pinMode(10, OUTPUT);    // set the SS pin as an output
   SPI.begin();            // initialize the SPI library
}
void loop() {
   digitalWrite(10, LOW); // set the SS pin LOW
   SPI.transfer(0x00);     // command to device to write at address 0x00
   SPI.transfer(44);       // send a new value to that address
   digitalWrite(10, HIGH);// set the SS pin HIGH
   delay(1000);
}
```

A huge number of sensors are available, often for less than $1 each. They will usually have a library available online which will make it easy to use from Arduino code, and examples of how to use it. There are also output devices available, like lights, displays, relays, and even motors. A great way to start is to buy a kit with an assortment of ∼30 such parts.

Lab: Arduino Lab

Download and install the Arduino IDE: https://www.arduino.cc/en/main/software.

Include your code (or at least the relevant parts of it) for each of the following, and show the behavior of the circuit.

1. Program an Arduino to blink a factor of 2 faster every 10 blinks, starting at 3 seconds per blink.

2. Read a digital input (e.g. a push button), using a pullup resistor on the input, and light an LED or not depending on the input. How do you choose the value of the resistor? What happens without the pullup resistor? Also try removing the external pullup resistor, but enabling the built-in one with pinMode(#,INPUT_PULLUP);

3. Read an analog input and write out an analog output (PWM) that scales with the input. Smooth the PWM output so there is less than 1% variation in the output, when input to a 1 MΩ load.

4. Read an analog voltage, react by moving a servo motor, and print out the angle using a serial connection on a computer. (If you don't have the servo motor, at least put the code there to do so, and explain how it would be wired.)

5. Send a character over software serial, capture the output on an oscilloscope, and explain the waveform seen. Beware to send a byte like Serial.write(1), not Serial.write("1"), which would actually send byte 49 in decimal according to the ASCII table.

Chapter 8

Printed Circuit Boards

Breadboards are great for quickly prototyping small circuits and experimenting. You can try things quickly and reversibly — when things don't work well, just unplug them and start over! Where possible, you should always prototype each small subcircuit of your project before making it more permanent. It will save you lots of time overall. For instance, if designing a device that takes in an analog signal from an electric guitar, pre-amplifies it, applies reverb processing, and then outputs it, you should try each part of that circuit separately on a breadboard. Input signals from a signal generator into each stage, and look at various intermediate signals and the output(s) using an oscilloscope. Use a signal generator as input *before* trying it with the guitar, by the way — you have much more control over the input frequency, amplitude, and output impedance. You could also build each piece on a separate breadboard and connect the output from each stage to the input of the next, thus testing them all together.

However, as your circuits become more complex, the breadboard becomes impractical. There are just too many connections going to too many places:

Even if each connection works 99.9% of the time, by the time you have 1000 wires, the chance at least one of them is not working is about 63% (and for me I'd say it's 100%)! And the chance that the circuit keeps working for more than a day is near zero, especially if you need to move it around a bit.

More Permanent Boards

One can solve some of these issues by using *perfboard*, an array of copper holes in which you can *solder* wires and other components. This is useful if you have a breadboard circuit you want to make more reliable and durable, at the cost of more challenging construction and far less ability to make later changes.

Any breadboard or perfboard circuit will also be quite big though, which adds to its impracticality. A modern device (like a computer motherboard) can easily have a million connections, which would require at least a 2000 × 2000 breadboard, which at 0.1" spacing would be 200"× 200", nearly 40 square feet.

And aside from just "messiness" issues, there are actual performance limitations to the breadboard (or perfboard). Each wire has self-inductance and parasitic capacitance, roughly proportional to the length of the wire. The breadboard (or perfboard) itself also has significant capacitance from the long rows of parallel connectors. As we've seen, these limit the bandwidth of the device, so high-frequency signals will be muted. And it is challenging to implement transmission lines on a breadboard, with controlled input and load impedance to match the characteristic impedance of the transmission line. You can get around these issues, for very small circuits, by soldering small components together directly, or with small wires, but it is challenging:

Lastly, we often want to produce more than one copy of a circuit. For instance, you may have one thousand channels of signals coming from a detector, each of which needs to be processed in the same way. We'd like to be able to automatically generate a thousand identical copies of our circuit.

PCB's

The *printed circuit board* (PCB) will solve all of these issues, and is indispensable to modern electronics. It will allow us to create high-quality, reliable, small, high-frequency capable, and reproducible circuits. And fortunately, thanks to modern manufacturing technology, PCBs can be produced quickly and are very inexpensive these days, in fact even less expensive than breadboards and wires. You can get 10 (identical, 2-layer, modest-sized) PCBs produced and delivered within a week (to most places in the world) for about $30.

A PCB is built from thin layers of copper, separated by insulating material called *prepreg*, usually made of a stiff fiberglass with resin such as FR-4, which makes it flame-retardant. There are also "flexible PCBs", which are much thinner and have a flexible plastic insulator instead, but we won't go into them here. You can get PCBs with various numbers of copper layers, with various different thicknesses of insulator between each layer, various dielectric constants and materials of the insulator, etc. The most common and cheapest is 1.6 mm thick and "2-layer", so a layer of copper on the top and bottom of the board. But for denser wiring, and to allow for more ground (or power) planes (discussed below), you often may need a 4-layer or even 6-layer PCB, which are still reasonably priced. Advanced high-speed, high-density, designs can require 20-layers or more! The copper layers are usually "1 oz", which means 1 oz of copper spread over 1 square foot of area, giving a thickness of about 1.37 *mil*, where a mil is one-thousandth of an inch. (Yes, PCB units are awful!) That's about 35 um, which is about half the thickness of a human hair — pretty thin! But recall that skin depth limits the usefulness of thicker conductors anyways at higher frequencies — at 10 MHz the skin depth is just 20 um.

The copper layers are what form the "wires" (called *traces* on a PCB) and "planes" of conductive material. To make the desired traces, a photo-lithography technique is used to remove the copper in each layer where it is not wanted, thus leaving copper traces. The copper is pre-coated with a protective layer that is sensitive to UV light. A mask with the circuit pattern is then laser-printed on a transparent sheet, and UV light is shone through this mask onto the coated copper. Where the UV light hits, the coating is degraded, exposing the copper. Then chemical etching is applied to remove the exposed copper, leaving the coated copper.

Vias

Of course, we also want to be able to connect a trace in one layer to a trace in another layer — otherwise two traces could never cross each other without touching. This is done using a *via*, a small, conductive, tube going from one layer down to another layer. They are made by creating a hole through the desired layers of the board, using a drill bit or laser, and then electroplating the surface of the hole with copper. The most common, and cheapest, type of via is the *through-hole* via, which goes through the whole board. That allows you to connect a trace in any layer to a trace in any other layer. However, all other traces in all other layers must go around the via or they will get connected as well. To avoid this, one may use the *blind* via that goes from an outer layer only to some inner layer, allowing other traces beneath it to still not be connected. The *buried* via can be used to connect only between inner layers. You can also have *in pad* vias these days, which are vias that are directly beneath, and connecting electrically to, the exposed pad for a component on the top or bottom of the PCB.

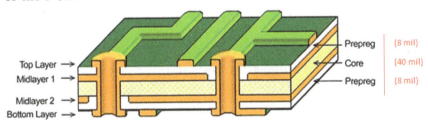

Soldermask and Silkscreen

Soldermask[1] is then applied to the outer layers, which protects the copper from oxidation. It is etched away from the pads that

---

[1]Usually the soldermask is green, giving PCBs their traditional green look, but other colors are also available these days. Red is nice, and black is uber-cool.

components will be soldered to. Those exposed pads are then coated with a more oxidation-resistant metal such as nickel, usually by dipping the board in liquid metal and then blowing it with hot air to a desired thickness, known as hot-air-solder-leveling (HASL). The soldermask is resistant to solder — it doesn't stick to it — whereas solder readily adheres to the exposed metal. This makes it easy to solder the components to the board without the solder making unwanted conductive connections. Finally, a silkscreen is printed on the top and bottom of the board (on top of the soldermask) with whatever text is wanted, such as the names and outlines of the components — so you know what to solder where!

Components

Most PCBs would be useless without the electrical devices and connectors, the *components*, that go on them. These can be soldered to the PCB in two ways (as discussed in detail later). *Through-hole* components have little wire "legs" that stick out of the component in precise places, which then go though through-hole vias in the PCB and are soldered in place. This is the component style used on breadboards (and perfboards) as well. *Surface-mount-devices* (SMD) use *surface-mount-technology* (SMT) to join pads on the bottom of the component to pads on the surface of the board. The through-hole connection is stronger, but SMT has many advantages. It takes up less board space since the holes can not be as small as the pads. SMT allows traces to pass under the pads in other layers without being connected to the leg of the device, whereas through-hole goes through all layers of the board. With SMT you can have components on both sides of the board directly below each other. SMT is also friendlier to automated soldering using pick-and-place and reflow techniques. Finally, SMT provides less parasitic inductance and capacitance for the parts since there are not wires between the part and the board. So, aside from connectors or other parts where mechanical strength is key, use SMT!

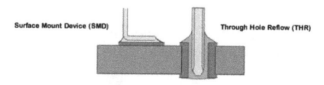

Surface Mount Device (SMD)   Through Hole Reflow (THR)

SMD parts come in many sizes. For instance 0603 means about 0.06 by 0.03 inches — yes, again, PCBs have awful units! Any smaller than this is not easily soldered by hand without a microscope.

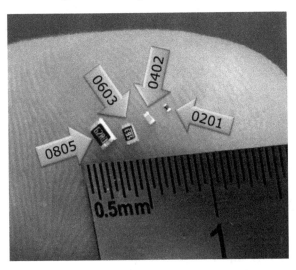

What types of components you need will be determined by what circuit or device you are making, of course. To find a specific component to use for a specific role on your device, one usually searches online catalogs from distributors, such as mouser.com or digikey.com. But for any type of component, you will often have many choices. For instance, a search on mouser for "0.1 uF capacitor 0603" gives 461 results at the moment for in-stock parts. Unless you have some special requirements, picking the cheapest, for the quantity you will need, will usually be fine. In this case I can get 100 such capacitors at 1.5 cents each, or $1.50 total. (The most expensive, in case you're curious, is $246 for 100, but can withstand 200 V and operate at up to 175 C!) Once you've settled on a component, you can also search across all parts distributors with engines such as octopart.com, for the best price or availability for a given quantity.

The collection of all components that will go onto the board is called a *bill of materials* (BOM). This will be a spreadsheet where each line is a part number, quantity of that part, the names of places on the schematic where that part will go, the price, etc. It will make assembly easier and cheaper if you minimize the number of different kinds of parts in the BOM — in other words minimize the length of the BOM. You don't want to have to search through lots of different parts if the same part can be reused instead. If you pay for assembly, they often charge extra per BOM item since the machine has to be reloaded with different parts. For instance, don't use 0.1 uF, 1 uF, and 4.7 uF capacitors if you can get acceptable performance using 1 uF everywhere. Don't use 50 Ω, 100 Ω, and 200 Ω resistors if you can get acceptable performance (and density) just putting 100 Ω resistors in series or parallel where needed.

PCB Design Software

To design a PCB, use one of the popular *electronic design automation* (EDA) programs. The basic steps, which we'll go through below, are:

(1) **Schematic capture**: This defines all the components and their "logical" electrical connections.
(2) **Layout**: This defines where each component "physically" goes on the board, and its orientation.
(3) **Routing**: This completes the logical connections, as defined in the schematic, with physical traces (and vias) between the components, as defined by the layout.

There are many choices for the EDA software, from free to VERY expensive ($10k+ per year license), with various features. As usual, different people will have their arguments for why their favorite one is the "obvious" best choice. Objectively, all the top ones are very good, and none of them are perfect. We will use Eagle[2] for the examples in this book. It is full of features, well-supported, and free for small 2-layer boards. Larger boards and more layers are free for non-commercial and educational use, which is great for scientists. Other ones to consider are KiCad,[3] which is fully open-source and free, and EasyEDA,[4] which is free, online, simpler, and closely connected to PCB manufacturing, parts, and assembly. Altium recently made

---

[2]https://www.autodesk.com/products/eagle/free-download.
[3]https://www.kicad.org/.
[4]https://easyeda.com/.

a free EDA tool called CircuitMaker.[5] But for serious professional work you'll want a subscription to Altium.[6]

We will *not* have room in this book to guide you through all the technical steps of using a particular EDA program for designing a PCB. There are countless other books and free online tutorials and videos available, which we'll make use of to design a PCB for this chapter's "lab". But we will outline the various steps and make some general points here about what will make a good design.

Schematics

The schematic defines your circuit or device. You will add components to your schematic from a *library*, wherever one is available for the component. Sometimes you need to google for it and download it, or generate it from the Mouser webpage. This will define the *pins* (or *pads* for SMD) of the component along with their names and draw them on a symbol for use in the schematic. It also will define a *footprint* for the component, which is the physical layout of the holes (or pads) that should be exposed on the PCB for the component to be soldered onto, for use during board layout (the following step). If it's a very new or rare component, you may need to create this component's library description (the pins, symbol, and footprint) yourself.

The schematic eventually has all the components shown and defines the electrical connections, or *nets*, that connect the components' pins or pads together. This should be run through an *electrical rules check* (ERC), which will make sure your circuit makes sense — for instance that all inputs and outputs from a component are connected to something, that nets all go from one place to at least one other place, etc.

In addition to its functional role, the schematic should also *explain* your circuit or device. Someone reading the schematic (including yourself, years later) should be able to understand how the circuit works, why the circuit was designed the way it was, and even why particular values for components were chosen. Following some simple conventions help achieve this:

---

[5]https://circuitmaker.com/.
[6]https://www.altium.com/.

- Group your components into logical, functional, blocks. For instance, your device might have a pre-amplifier stage, a digitization stage, and a digital processor. Each of those units should be in its own area of the schematic, clearly showing the input(s) and output(s) to and from the block. That usually includes showing decoupling capacitors nearby to the points they are meant to physically decouple.[7]

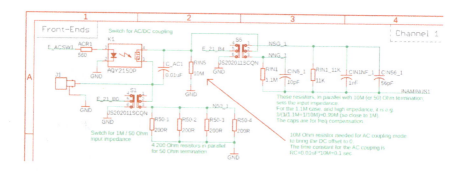

- The signal "flow" should go from left to right, so inputs for each block should be on the left and outputs on the right, where possible. Earlier blocks in the signal processing chain should likewise be on the left, where possible, and feed into later blocks on the right. For instance, the pre-amplifier block, which takes inputs from the outside world, should be the leftmost and "first" block. Of course, for larger schematics, you will run out of space going left to right. Just start going down, or to following pages. The schematic need not fit on one page. Use multiple pages and give them sensible names.
- Voltage should be higher at the top of a block and lower at the bottom. So (positive) voltage power inputs should be at the top and ground should be at the bottom.

---

[7]Sometimes it is better to put all decoupling capacitors on a page (or pages) at the end of the schematic for greater overall readability, particularly for large digital devices which could require hundreds of capacitors. In that case, at least make clear what component the capacitors are meant to be near to.

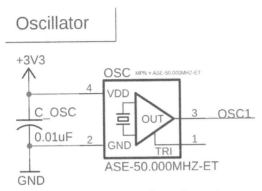

- Give nets sensible names, especially when they are carrying an important signal, and show the name on the schematic.
- Show all names and values of all components, and part numbers where important. We don't likely care what brand of resistor is being used, but do show what op-amp was chosen, since different op-amps can have very different specifications.
- Component names should follow conventions. For instance, resistors start with "R", like R_1, R_2, etc. Capacitors are C_1, C_2, etc. Inductors are L_1, L_2, etc. Diodes are D_1, D_2, etc.
- Add text explaining design choices and what each part of the circuit does.
- An excellent example is the Arduino schematic: https://www.alla boutcircuits.com/uploads/articles/Arduino_schematic.png.

Layout

Once you have completed the schematic, all the footprints of all the components will be shown (in another window), along with *airwires* showing which pins need to be connected to which other pins according to the schematic. The game is then to move, rotate, and flip (to the other side of the board) each component so that it will be in a "sensible" location. You also need a *board outline* that defines the physical size and shape of the board. Be sure to show the proper component names and outlines in the right places on the top and bottom silkscreen layers of the board.

Don't forget to add *test pads*, small copper circles on the board surface connected to important or difficult to access pins (or pads) on your components, so that you can test and debug your board later. It's also nice to design in some good spots for the multimeter or oscilloscope ground probe to attach to, like a set of ground pads or holes near the edge of the board. For higher-frequency signal test pads, have a ground pad nearby for a low-inductance oscilloscope ground spring.

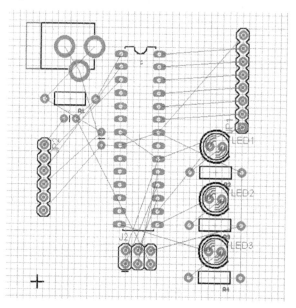

As we are designing a physical object (the PCB) that must be actually manufactured, it is important to tell the software about the manufacturing capabilities and tolerances of the PCB manufacturer. For instance, what is the minimum trace width? What is the minimum distance between traces? What is the minimum diameter via hole? These myriad parameters are incorporated into a *design rules* file that should be available from your PCB manufacturer for your EDA software, which should then be loaded. The software will then obey those rules automatically to the extent possible during layout and routing. A *design rules check* (DRC) can be performed at any point to see all violations of the rules that you need to fix.

It's best to keep all components on the top of the board, but if you need more space, it's fine to put some on the bottom of the board as well. This makes soldering and assembly (especially automated assembly) a bit more challenging and expensive, but it's not a big problem these days. It also makes testing and debugging a bit easier if you can reach all the components at once.

In an ideal world, it wouldn't matter where components were placed — the functionality of the device would just be defined by the connections on the schematic. Of course, connectors for inputs and outputs need to be moved to where they physically need to be on the board, to fit in some case, or connect to another board, or just make them accessible. This usually means putting them on the edge of the board. And don't forget that decoupling capacitors need to be placed near to the pins (power pins, usually) of the components they are providing decoupling for. The lower the inductance of the connection, the higher the frequency that can be decoupled.

Another real-life constraint on component layout will come from the next step, routing. If the components are randomly put all over the place, there will be many long connections, with many traces needing to cross over other traces without connecting. This will require many more vias and much more space and possibly even more layers (thus more cost) than if components are moved and rotated so as to minimize the length of traces and the need for traces to cross. So, the first step is to simply move components so they are physically in similar block areas, like they are in the schematic. For instance, the components that are part of the pre-amplifier block in the schematic would all go into the same area on the board.[8] Having components from the same functional block near each other will also make the board easier to understand and debug when you (or someone else) are looking at it on their desk. It's far easier to find the right component to probe with the multimeter or scope when the component is localized to an area of the board dedicated to the logical function of the circuit being studied. You should also try to arrange and rotate the components so they minimize trace crossings along with trace lengths, as this will make the routing step easier.

As you'll see, you won't get your layout right the first time. Once you start routing the board, you will want to adjust the layout a bit. It's normal to iterate between layout and routing, adjusting the

---

[8]It can be helpful to do the layout of each block just after the components of a block are added to the schematic, rather than sifting through all the components once the whole schematic is done.

layout of a section of the board once you've learned something about how the routing will look, and then re-routing.

Ground Plane

The real fun in layout and routing is arranging for good signal integrity and minimizing noise and EM radiation.[9] If you have any "high-speed" signals on your board, above 1 MHz or so (or rise/fall time below $0.35/1\,\text{MHz} = 350\,\text{ns}$), you should include a *ground plane*. Note that nearly all digital components these days meet this definition of high-speed. The switching on even cheap microcontrollers can have rise/fall times of a few ns or less! The ground plane is a solid layer of copper connected to "ground". It is essential for dealing with the issues of high-frequency signals we discussed in the last chapter.

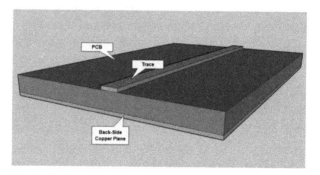

The ground plane forms a path with the lowest possible inductance between all points on the board. This is critical since (as discussed in the last chapter) the voltage difference between two "ground" points on the board will be given $V = IZ$, where I is the current flowing between the points, and Z is the impedance of the path. It is primarily the inductance (not resistance) that determines the impedance of the path at high frequency, since the inductive reactance rises with frequency. This ground plane will have a maximum inductance between two spread out points of about 1 nH, so $Z = i * 2\pi * 0.2 * 10^9\,\text{Hz} * 10^{-9}\,\text{nH} = {\sim}1i\,\Omega$ at 200 MHz. For a 50 mA current, typically near the maximum for a switching digital

---

[9]Countries have strict rules about how much *electromagnetic interference* (EMI) the board is allowed to radiate in particular frequency ranges while operating.

output, this would lead to a voltage difference of 50 mV, which is usually acceptable. A wire connecting the two points could have 20 times more inductance, leading to an unacceptable 1 V difference during switching.

Another advantage of the ground plane is to avoid ground loops. Magnetic fields changing their flux through such a loop would induce voltage around the loop. The ground plane thus helps to shield the board from outside EM radiation and nearby EM fields, as well as keeping the board from radiating and causing interference for other devices.

But most important of all, the ground plane provides a way for us to make good transmission lines in the PCB, and to keep the signals in those transmission lines reasonably isolated from each other, if we're careful. Recall that we need to consider transmission line effects whenever the wavelength of the signal is about 10 times the length of the trace, or less. Beware that the wavelength in PCB is about 2 times smaller (and the propagation speed thus 2× smaller) than in air, due to the dielectric of the PCB substrate. For a 15 cm trace, that means $\lambda = c/f$ is around or below 3 m, which is when f is above $c/\lambda = 3 * 10^8 \, \text{m/s}/3\,\text{m} = 100\,\text{MHz}$. From the last chapter we saw that it is really the risetime (or falltime) of the signal that determines the highest frequencies we need to consider. For a 1 ns risetime, we need to consider frequencies up to about $0.35/1\,\text{ns} = 350\,\text{MHz}$, which become important to treat as transmission lines for traces longer than about 4 cm! Another way to understand this is that signals travel at about half the speed of light, or 15 cm/ns in a PCB, so transmission line effects will be important whenever the length of the rise or fall time of a signal in the PCB is comparable to the length of the trace. This is another strong reason to arrange components so as to keep traces short — if you can keep them short enough (for their given rise/fall time), you can ignore transmission line effects.

Even with a ground plane, our PCB traces will induce voltages and currents in each other. Digital signal traces, with their high frequency components and thus large dV/dt and dI/dt will make especially large changing electric and magnetic fields that can make significant voltages and currents in other traces. We'll look at this in detail below, but for now we'll just mention that analog signals will generally have a much smaller tolerance for this induced noise than digital signals. (This is one of the main advantages of digital signaling!) For instance, you may be very concerned about an additional

50 mV of noise on an analog input to an op-amp in some part of your circuit, but you will not care about that level of noise on a 3.3 V digital signal since it is never enough to change a 0 to a 1 or a 1 to a 0. It's thus critical during layout to keep analog components away from digital components, wherever possible, to keep any digital signals from passing near any sensitive analog signals. A good rule of thumb (validated by our discussion of transmission lines on PCBs below) is to keep analog and digital components (and the traces going to them) away from each other by at least 20 times more than the separation between the components and the ground plane beneath them (H).

Of course sometimes a digital signal will need to talk to one of the analog components. Usually you can afford to simply slow down the rising and falling edges of the digital signal, using a low-pass filter (made by putting a ~1 kΩ resistor in series with the signal on the digital side). But if the signal must be fast, you must be careful that the digital signal is isolated from the analog signals, by using generous spacing between it and any analog tracks, and even reserving a layer of the board on the analog side for digital signals, shielded by ground planes from any analog signals. Some components are both analog and digital however, such as an *analog to digital converter* (ADC). Fortunately, the designers of such parts are almost always kind enough to put analog pins on one side of the part and digital on the other. Place the part between the analog and digital sections of your board and rotate the part so that the analog side is pointing towards the analog section of your board and digital pins towards the digital section of your board.

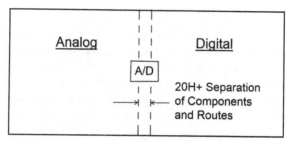

Keep EVERYTHING Analog in Analog Section!
Keep EVERYTHING Digital in Digital Section!

When designing a PCB, it is easy to forget that signals going from one place to another are only *half* of the current path. Current must also flow back from the destination to the source, through the ground connection! This is called the *return current*. One downside of using a solid ground plane in a PCB is that all signals will be sharing the same ground conductor for the return current. This can lead to "ground bounce" problems, as discussed before, since the return current from one signal will cause a voltage (across the ground impedance of the return path) in the "ground" of other signals. The effect will appear as "noise" on the other signals. Fortunately, as we'll see in the next section, the return current likes to take paths of low impedance, which means low *inductance*, or low *loop area*, at higher frequencies (above $\sim$100 kHz or so on a typical PCB). So higher frequency return currents will be isolated to areas nearby (directly under or above) their signal paths. But lower frequency return currents will spread out over the board more. In other words, the "ground" for a low-frequency signal will "see" a large area of the board around its signal trace. Also fortunate though, is that the impedance of the ground plane at these low frequencies is quite low, so the voltage noise induced is small. But it's good to consider during layout that low frequency return currents will be spreading out over the ground plane. Don't place sensitive (analog) components in the "path" of large return currents.

You will sometimes hear someone (or even app notes or data sheets!) advise you to *split* the ground plane into two (or more) sections, for sensitive analog on one and digital on another, for instance. You should generally ignore that advice! It will usually cause more problems than it will solve. For instance, the two halves of the ground plane will form a nice dipole antenna and couple nicely to EM radiation in the 100 MHz range and above. You will be receiving noise as well as radiating it out, whenever you have anything sensitive to or generating such frequencies on your board, which is nearly always the case these days. If you are careful about considering where return currents will flow on your PCB, and keep analog parts well-separated from digital signals and return currents, you will usually have no trouble with a solid ground plane.[10] An exception is when

---

[10] https://www.scribd.com/document/835303172/26-2-June2001pcd-Mixedsignal.

you have very sensitive analog components (e.g. 16 bit ADCs or higher), and/or when frequencies of interest are below ~10 kHz. In that case, you can consider, if really needed, putting *slots* in the ground plane separating the analog and digital sections, except under the ADC or whatever small region connects the analog to digital sections. This will ensure that low-frequency return currents from the digital side do not flow through the analog side, as long as no digital traces go into the analog section. You can also use *star grounding,* where a separate ground trace is run to each component from a common ground point. Another exception is if you have high-current power supply components on the same board as other components. Those are often better to isolate with their own ground plane, or better yet just input clean DC power from an external adapter.

Stackup

Now that the components are roughly in the right places, it's time to replace the airwires with actual traces. Since we have a ground plane on the bottom of the board, the traces should all be confined to the top layer. Of course, a trace may have to cross another trace, and this is impossible if they are both always in the same top layer. So where necessary, a trace can drop down to the bottom layer using a via for a short bit, go under the other trace, and then back up through a via to the top layer. The goal is to keep the ground plane as intact as possible, especially below high-frequency signals where transmission line effects will be important, as discussed below. But small breaks in the ground plane beneath lower-frequency signals are OK.

Once you have lots of signals, need dense connections, or need transmission lines to cross each other, two layers will not suffice. A 4-layer board or even 6-layer board is needed — but don't worry — they're still very affordable. These give significantly more routing space and should be all you need, unless you are working with very dense and high-end parts like FPGAs or CPUs. With more than two layers, we need to consider the *stackup*, which is the spacings between the layers and what each layer will be used for. On a 4-layer board, you may be tempted to keep just one ground plane, on the bottom say, and use the other three layers for traces. That will be bad for higher-frequency signals, as we'll see below, since the ground plane will be far away from some of the traces. The copper layers of a multi-layer PCB are usually not evenly spaced — they come in pairs of closely spaced layers, with the pairs separated by larger layers of "core" prepreg material. A much better stackup is shown below, with *two* ground planes on inner layers, and two signal trace layers, on the top and bottom. This will allow for far better treatment of transmission line signals, maintaining impedance, as discussed in detail as follows.

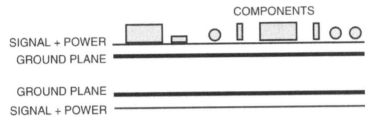

For six layers, you can usually afford to dedicate one of the layers to a *power plane*. This would be made of solid copper sections, each connected to a voltage different from ground. The plane will form a capacitor with the neighboring ground plane, with nominal capacitance (a few pF maybe), but very low inductance. It can deliver power on extremely small time scales and decouple very high-frequency (GHz and above) fluctuations in the voltage supplies. You still should use two additional ground planes, leaving three layers for signal traces, each of which is still adjacent to a closely spaced ground plane.

Routing

The last part of the board design is to make all the needed electrical connections between pins of components. The airwires show you what connections need to be made, according to your schematic. Most modern EDA software will come with an *auto router*, which will attempt to route all unrouted airwires (or a selected subset of them) for you automatically. This is usually fine to use for very simple boards, or simple sections of a board, but not for cases where transmission line effects or noise or crosstalk are relevant. So do important traces (transmission lines, sensitive analog lines, etc.) first, by hand, and then you can let the auto-router do other simpler traces. But then check the auto-routed traces carefully to see where they go.

Also be careful about the widths of the traces. Narrow ($\sim$8 mil) traces are fine for low current signals, but power traces carrying significant current may need much wider traces. Use an online trace width calculator to determine the minimum trace width needed for a given current, length, and acceptable voltage drop.

Often you will have a component with several pins that can be used interchangeably. For instance, a dual op-amp will have two sets of inputs and output, or an FPGA may have hundreds of pins and you can program which pin does what. But you may find during routing that traces would not have to cross, or traces could be shorter, if the pins were in a certain order. You can use *pin-swapping* to simply redefine which pins of the component do what (on the schematic and thus in the layout as well), and then perform the newly simplified routing. Rotating and moving around components will often help a lot as well.

We've been careful during layout to keep analog components separate from digital (by at least 20 H), but we must maintain that discipline during routing, making sure that traces carrying analog signals don't get within 20 H of those carrying digital signals. Remember to slow down a digital signal (using a resistor, as discussed above), and maybe even keep it shielded by a ground plane from analog traces, when it must venture into an analog region (and never route it into the analog section at all if you took the risky step of adding slots to the ground plane).

Routing Transmission Lines

It's critical to be conscious of which signals on your board will need to be considered as transmission lines, due to the length they need to be and the rise or fall times they need to carry. Fortunately, making transmission lines on a PCB is easy — in fact all the traces are automatically transmission lines, and pretty good ones even, as long as they are over a ground plane. We just need to be careful to control the impedance of the line and its termination. A trace on the top layer, separated by dielectric, with a ground plane on the layer below, is called a *microstrip*. The signal current travels in the trace and the return current in the ground plane. Displacement current flows between the trace and the ground plane wherever the voltage difference between the trace and the ground plane is changing. The characteristic impedance of the microstrip is given by the width of the trace, the thickness of the trace (0.035 mm or 1.37 mil for standard 1 oz copper), the distance between it and the ground plane (H), and the dielectric of the board material between it and the ground plane (and the dielectric of the air above it, but that is very close to 1). The dielectric constant of FR-4 is about 4, and H is about 0.1 mm or 4 mil for 4-layer or 6-layer boards — but you have to check with your PCB manufacturer.[11] You can find microstrip trace impedance calculators online, but a good rule of thumb is that the impedance is about $50\,\Omega$ for trace width of $2 * H$ (so about 8 mil for $H = 4$ mil). Unless you have some special reasons, use the standard $50\,\Omega$ impedance for your traces.[12] To terminate the microstrip, just add a 50 Ω resistor (or 47 Ω, which is easier to find) between the trace and ground, at the end of the trace. It's a good idea to consider matching the source impedance to that of the line as well, since then any reflections from the end of the line (or noise) will get absorbed efficiently at the start of the line. Often the source will already have 50 Ω impedance (check the datasheet of the part), but if not you can add a resistor in series with the trace, at the start of the trace, to raise the impedance up to 50 Ω.[13]

---

[11]For instance, https://jlcpcb.com/impedance.

[12]Sometimes you'll need a trace thinner than 8 mil, but the impedance doesn't change too fast. It's only about 50% higher (75 Ω) for half the thickness (4 mil), which is about as thin as inexpensive PCB manufacturers support.

[13]It is rare that the source impedance of a part would natively be larger than 50 Ω.

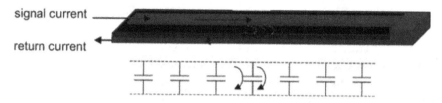

signal current

return current

Please notice a very important point about transmission lines — the signal energy does not flow in the signal wire — it flows *between* the signal wire and the return path! Fortunately, for high-frequency signals (above ~100 kHz or so for a typical PCB), which includes those in transmission lines on a PCB, the return current, and thus the energy in the EM fields for the signal, is isolated to a small region under the signal trace, assuming it is over a continuous ground plane. The reason can be thought of this way: the return current likes to take the path of lowest impedance, but for high frequency, the impedance is dominated by the reactance of the inductance of the path, not the resistance of the path, and the inductance is minimized when the loop area of the return path (and signal path) is smallest, which is when the return path is directly under the signal trace. For the return current path to be directly under the trace, we of course need the solid ground plane. If there is a gap in the plane, the current must travel around the gap (except for the part of it that can travel as displacement current across the capacitance present in the gap), significantly raising the inductance of the path (leading to more crosstalk, as discussed below) and disturbing the impedance of the line, leading to reflections. Lower-frequency signals (below ~100 kHz) have the ground return current spread out more, and a large part of the whole board at DC, so we need to be concerned about where these go, especially near sensitive analog components, as discussed above.

For greater noise immunity (especially reduction of far-end crosstalk, discussed below), an option is to put the trace inside the board, between two reference (usually ground) planes. This is called the *stripline*. It's an even better transmission line, more like a coaxial cable, with ground surrounding the signal (well, at least on the top and bottom, but since the board is so thin it doesn't matter much that the sides are "open"). The drawback is that you need more ground planes, so more layers, at least 6, which costs a bit more. Stripline traces will have a wave propagation speed about 15% slower than microstrips, since the effective dielectric constant is larger (no air on one side). The characteristic impedance of stripline is about 0.6 times that of microstrip, for the same parameters (trace width, H, dielectric, copper thickness).

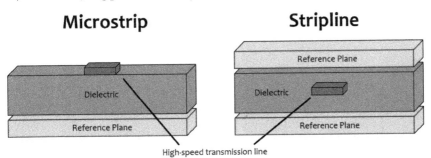

High-speed transmission line

Ideally, the signals going down one transmission line would not interfere with or leak into other traces on the PCB. But for traces on a PCB the EM fields are not completely shielded by surrounding ground and thus do spread out some. These (changing) fields will induce voltage on nearby traces, especially when they are parallel to each other. This leakage of one signal on the board (the *aggressor*) onto another net (the *victim*), is called *crosstalk*. Imagine an aggressor line running parallel[14] to a nearby victim line, over the same ground plane, and then launching a signal rising edge onto the aggressor. Capacitive coupled crosstalk voltage in the victim is positive (the same polarity as the signal voltage relative to ground) moves forward, along with the signal on the aggressor, but also backwards on the victim line.[15] It is a fraction of the signal voltage given by the amount of capacitive coupling between the two lines. Forward crosstalk builds up into one big pulse that looks similar to the signal pulse and arrives at the same time as the signal pulse, since the victim crosstalk is moving the same way as the aggressor and just continues to add. But backwards crosstalk will continuously be generated but have farther and farther to travel back to the near of the victim line.

---

[14] Traces that are perpendicular to each other will have a small region of overlap, and that region will have electric fields induced in the victim that are not in the direction of the wire, so will induce very little voltage in the victim.

[15] https://www.youtube.com/watch?v=EF7SxgcDfCo; https://practicalee.com/crosstalk/.

It will thus be small and constant for twice the length of time it takes the signal to travel down the aggressor line.

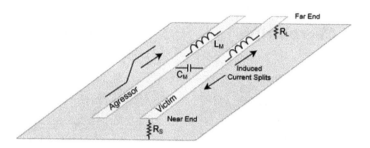

There is also inductively coupled crosstalk that moves forward and backward on the victim.

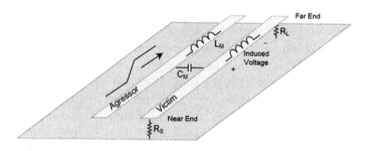

But the forward inductive crosstalk, which also builds up into one big pulse, is *negative* in polarity to the aggressor voltage, and thus will *subtract* from the capacitive forward crosstalk. The backwards inductive crosstalk is still positive. The inductive coupling is usually bigger than the capacitive coupling for two microstrips, so there is a net negative *far end crosstalk* (FEXT) pulse that arrives at the same time (and for the same length of time) as the aggressor signal. For two striplines, the capacitive and inductive couplings are equal, since there is uniform dielectric between them, so there is *no* far end crosstalk! The *near end crosstalk* (NEXT) adds between capacitive and inductive, for both microstrip and stripline geometries, and is still twice the length of time it takes the signal to travel down the aggressor line, but it is small since it just continuously dribbles back and doesn't add up as the signal is traveling.

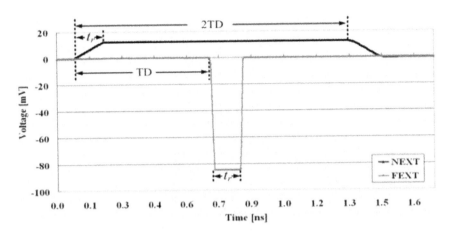

Clearly, we want to keep near and far end crosstalk below acceptable levels for each combination of traces on the PCB. Some traces, such as more sensitive analog signals will demand greater isolation, while digital signals may be more robust. But too much crosstalk can cause logic errors even on digital buses. Aside from using striplines to cancel FEXT, to reduce crosstalk one must reduce the capacitive and inductive coupling. A good rule of thumb is to separate microstrips by at least 5H to reduce crosstalk below about 1% of the aggressor voltage, assuming typical ~1 ns rise (or fall) time and 50 Ω lines on a PCB. (Recall that H is the distance between the trace and the ground plane, i.e. the thickness of the dielectric separating them.) For a 50 Ω microstrip trace of width 2H, that means separating the tracks by about 3 times the width (so you may sometimes hear of it as the "3W" rule). Crosstalk is also proportional to the distance during which the traces are running parallel to each other, and FEXT is inversely proportional to the rise (or fall) time. For instance, a 10 inch pair of parallel 50 Ω microstrips just one track width apart with a rise time of 0.3 ns could have a FEXT of more than 15% (!), but reducing the length to 2 inches and lengthening the rise time to 0.5 ns and separating them by twice the track width reduces the FEXT to just 1%. Of course the best is to move to striplines to nearly eliminate FEXT, but that requires more board layers and care.

You may hear someone suggest (or think on your own) of adding *guard* ground traces or *pours* (solid regions of copper filling areas between traces) between two signal traces, to reduce crosstalk. They don't do much, since the fields extend through the dielectric or air, and the guard traces or pours are very thin. If you use them, make sure they're very well grounded — use ground vias in the guard whose spacings are less than half the wavelength in the PCB of the highest frequency you care about, otherwise the guards will resonate at some frequencies you care about. It is far easier to just not use them.

So making transmission lines in PCBs is easy, but how do you change layers with them, to route them across other signals? We can use a via, as before, to switch the signal trace from one layer to another. But we now have to be careful not to destroy the transmission line properties of that signal when doing so. Otherwise the impedance of the transmission line will change at that point, and cause reflections, larger inductance and crosstalk, etc. We need to give the return current, which was flowing in the ground plane just beneath the signal, an easy path to switch the ground plane (or planes, in the case of a stripline) to a new plane (or planes) next to the new signal layer as well! The trick is to place a second (ground) via just next to the signal via, connecting the two (ground) planes.

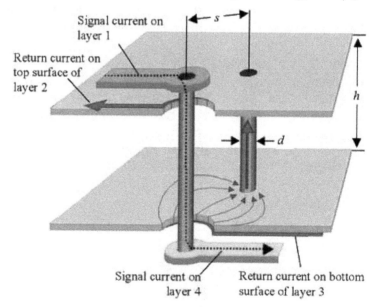

Signal current on layer 1

Return current on top surface of layer 2

Signal current on layer 4

Return current on bottom surface of layer 3

You don't need this extra via unless you're changing the ground plane that will be carrying the return current. Changing which *side* of a plane is carrying the return current does not need the extra ground via (though it doesn't hurt either). If you're going from layer

1 to 3 (on a 6-layer board, for instance) with a ground plane on layer 2, the skin effect will restrict the return current to the top of layer 2 when the signal is on layer 1 and the bottom of layer 2 when the signal is on layer 3. But when the signal switches from layer 1 to 3, the return current can flow through the via hole in layer 2, which connects the top and bottom of layer 2 at the hole edge (as well as making a separate conductive surface for the signal).

Be careful to couple as weakly as possible to traces or parts on the board that will be connected to cables. Those cables will be good antennae at some frequencies (e.g. half-wavelength as discussed in the previous chapter). EM energy will be nicely radiated to (causing EMI issues) and received from (causing noise) the outside world from / to nets connected to cables. It's thus critical to isolate those nets from other nets on your board that will have or be susceptible to "high-frequency" (half-wavelength of the longest possible cable length/c) energy. To be safe, use stripline traces for the most sensitive traces, since coupling to the outside world (and other traces) is smaller for them than for microstrip. Even better is *differential* stripline traces, such as LVDS (discussed later).

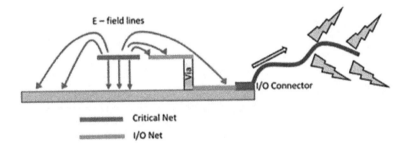

Ordering

Once satisfied that the layout and routing is complete, and the DRC passes with no errors, it's time to order the actual PCBs from the manufacturer. There are many inexpensive PCB manufacturers available with online web interfaces. The standard way to send your board design to the PCB manufacturer is by using an antiquated file format called *gerber files*. This is a set of files, each describing a layer of your board, or silkscreen, or drill holes, etc. Search online for how to export gerber files for your EDA software. They are then zipped together into a zip file, and then you can upload that zip file to the PCB manufacturer via their web interface. It should automatically detect the size of the board and the number of (copper) layers. Remember to select your board stackup here in the "controlled impedance" section, if you used transmission lines in your design assuming certain board stackup dimensions. There are a few other options, like colors etc., and defaults should be fine. You will get an automatic instant price quote, can then add the board to your cart, select shipping, and pay.

Assembly and Soldering

Once your PCBs arrive, it's time to *solder* (pronounced "säder", sort of rhymes with "water" in my New York accent) on the components. Both through-hole and SMD components are soldered on the PCB, using the same solder, but with slightly different techniques. Various types of solder are available, but for small-volume production it is fine to use *leaded* solder, which is an alloy of ~60% tin and ~40%

lead. It is much easier to work with than *lead-free* solder, which melts at a higher temperature, amongst other issues.

The way solder works is by melting and then *wetting* the surfaces of the two metal areas to be joined. Wetting is where the molten solder dissolves and mixes into the (surface of) the other metal. This forms an *intermetallic bond* between the solder and the other metal. The other metal does not melt. Copper, for instance, has a melting point of over 1000°C, whereas leaded solder melts around 200°C. The other metal just has to be hot enough to allow the solder to dissolve (wet) its surface.

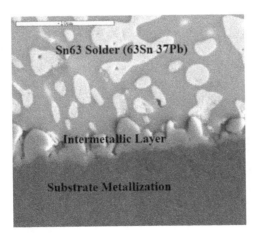

The main complication in soldering (in Earth's atmosphere) is that most metals, such as copper, very quickly form a metal-oxide layer on their surface that prevents wetting and bonding. It's therefore necessary to use *flux*, an acidic liquid that removes the metal-oxide layer and seals it from the air to prevent new oxidation. The flux also modifies the surface tension of the solder and helps it to "flow" onto bare metal, and only onto bare metal, e.g. not on the soldermask. The surface tension can be large enough to even align parts on the board. Flux can be added by hand as a paste or gel or from a liquid pen. It often comes integrated into the middle of the strand of *rosin-core* solder used for soldering. Isopropyl alcohol (99% concentration, from the drug store) should also be used to pre-clean oils and dirt from any metal being soldered. It's also good practice to clean any flux left over from soldering off the board (again using the alcohol or a specialized *flux cleaner*). Over time, the flux will attract moisture from the air, which can corrode the board and solder joints.

 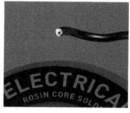 

We'll practice some simple soldering of through-hole and SMD parts on your PCBs. For advanced ICs with tens or hundreds (or thousands!) of pins or pads, typically in a *ball grid array* (BGA) package, a slightly more advanced method is used. Solder *paste* or *balls* are applied to the chip using a stencil, and hot air or an oven is used to melt the solder and make the connections to the board. For higher volume (more than 5 or 10 boards), or a complicated board, you likely need automated PCB assembly (PCBA), by a company like macrofab.com. Or you can contact your PCB manufacturer to see if they will do PCBA for you after making your PCBs. You can either send them your components, or use *turn key* service where they order the parts for you.

Here are a few other soldering tips (no pun intended!). You'll need more power from your soldering iron to get hot big components (or those connected to ground planes or other big pieces of metal), since they have large thermal capacity. This means that the "ground" pins of components can require more heat than others, since they are thermally connected to the ground plane. It may be necessary to use an *infrared board preheater* to heat the whole board to ~150°C to make soldering possible. Sensitive plastic connectors can melt if heat is applied to them during soldering. It may be necessary to place the component on the board and then heat the board (with hot air) from the other side, so called *bottom heating*. For removing solder, use *desoldering braid*. Keep your soldering iron tip clean using a metal mesh.

Lab

The exercise related to this chapter is a multi-week Final Project, where you design, layout, assemble, debug, and test a PCB for an electronics project of your choice.

Week 1: PCB schematic and components

1. Decide on a suitable project. It should not be too advanced. There should be no more than a few "significant" components, like complicated IC's.
2. Make sure that all components you want to use are available and in stock.
3. Read the datasheets for all the IC's involved, and perform any needed circuit simulations or calculations.
4. Use EDA software to design a schematic for the project.

Week 2: PCB layout and ordering

1. Choose a stackup for your PCB. In most cases this will be just 2 layers, a top and bottom.
2. Use EDA software to layout the components on the PCB.
3. Form a ground plane on the bottom layer. Most traces will then be on the top layer, along with all components. (Traces can drop down to the bottom layer briefly to "cross" other traces on the top layer).
4. Route the traces on the board.

5. Order the PCB from an online quick-turn manufacturer (e.g. JLCPCB).
6. Order your components from an online distributor (e.g. Mouser).

While waiting for PCBs and components to arrive, move on to Chapter 9 (FPGAs) and do the Chapter 9 Lab.

Week 3: Soldering and debugging

Now that the PCB and components have arrived, we will assemble the PCB and try it out! Recall the basic assembly steps:

- Soldering of through-hole components:
  - Rip off ~1 ft length of solder wire
  - Put leads through holes, flip board over
  - Heat hole and lead with iron, then touch solder to them, let it flow it (about 1 second), then remove iron
  - Make sure component doesn't move before solder cools
  - Clip leads with edge cutting pliers
- Soldering of surface mount (SMD) components:
  - Put a little solder on one pad with iron
  - Heat solder with iron, then with other hand use tweezers to grab and place component, then remove iron
  - Make sure component doesn't move before solder cools
  - Heat next pad and component lead, add solder, let flow, remove iron, let cool

Once assembled, power on your PCB and cross your fingers! Use a power supply with a current limiter to power the board. This way, if there's a short, it won't damage your board or a component on the board. Once it's shown to be stable at an acceptable current, you can then move to a standard power source.

Test basic features of your board and then debug things that don't work as expected. (There always are some — don't feel too bad!)

- Are the various voltage power rails at the voltages you expect?
- Is the current being used sensible?
- Are the input signals as expected?
- Look at the output signal(s) of the first stage of the circuit — are they as expected given the input? What about for other inputs?
- How about for the following circuit stages?

Week 4: Testing and documentation

Hopefully by now the board more or less functions as expected. Characterize its performance, comparing it to the expectations from theory and simulations where possible. Imagine you're creating a datasheet describing your project.

- Describe the main features of your project, its purpose and goals.
- What are the main specifications of your project — input voltage and frequency ranges, how to power it (voltages and currents needed), output characteristics?
- Make several plots showing the performance of your project.
- Discuss the qualitative limitations of your project (e.g. only two ranges).
- Quantify the noise or other imperfections in your device's output(s).

Week 5: Final project presentation

Present to the class a 10-minute slide presentation explaining the final project, the PCB design, operation of the circuit, and results (compared to theory and simulations where possible).

Chapter 9

FPGAs

Let's say you want to do some relatively simple digital logic. For instance, you have three detectors, A B C, and you want to count how many times A and B make a detection, while C does *not* detect anything. (C would be called a "veto".) In the old days, you would use digital logic chips to do this, as shown below. First, a "not" gate would invert C. Then an "and" gate would take in A B. Then another "and" gate would take in A&B and ~C, and output the AND of them. A logic diagram of such a process is shown below. This output could be fed into a counter chip.

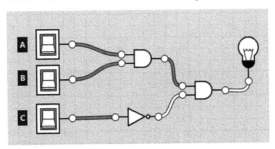

You could still do this of course, and for very simple logic like this you still might. It is inexpensive and easy. But requirements have a way of increasing, and designs that start simple have a way of becoming more complicated over time. You may want to add more detectors. Or you may want to place timing requirements on the detections, like requiring A and B to be on for at least 80 ns, or C to not fire within 130 ns after A and B. This expanded functionally

155

becomes unwieldy quickly. And a new hardware design would be constantly needed.

The FPGA

A microcontroller would not work well. First, they're too slow. They have typical clock speeds of 16 MHz or so, but we need to constantly update the output every ns or so. They are also not "real-time", meaning that interrupts or other work the CPU needs to perform will alter the timing of the output updating in unpredictable ways. You don't want your output to depend on how your debug printouts are being sent to a computer, for instance. It would also be very inefficient and overly complicated, using millions of transistors to perform the same job that could be done with just a few.

A *field programmable gate array* (FPGA) combines the best of both worlds for tasks like this, giving the flexibility of re-programmability using software and the low-level efficient hardware use and speed of discrete logic chips. Input/Output Blocks (IOB) on the outside of the chip handle interfacing to/from electrical signals on the wires attached to the chip. Analog signals coming in on the wires are digitized into 0's and 1's by the IOBs. The IOBs can also take 0's and 1's from the FPGA and output them as analog signals on the wires.

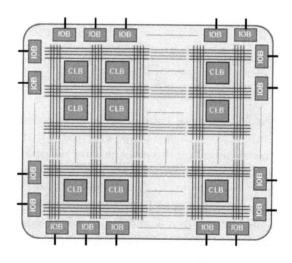

Inside the FPGA, each Combinatorial Logic Block (CLB, or sometimes just Logic Block (LB)), can take digital signals as input and perform logic operations on them, producing output(s). Which logic operation is performed is determined by the state of the LookUp Table (LUT). The LUT can be programmed to be an AND gate, OR gate, NOT gate, etc. — anything that takes the right number of inputs and gives the right number of outputs. The LB also has a flip flop, which is basically a 1-bit memory register. This allows the output of the LB to also depend on the previous output of the LB, allowing for far more complicated and useful behavior.

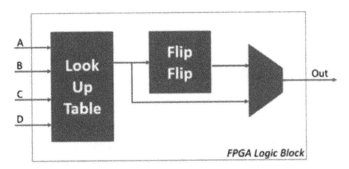

Hardware Description Language

FPGAs are programmed using a *Hardware Description Language* (HDL). There are two popular such languages, Verilog and VHDL. For example, in verilog we could have the following "module" which defines an AND gate:

```
module and_gate (input1,input2,output1);
    input   input1;
    input   input2;
    output output1;

    wire    and_result;
    assign and_result = input1 & input2;
    assign output1 = and_result;
endmodule
```

The module has a name, and the wires coming into and out of it are named. We then say which wires are inputs and which are outputs. These inputs and outputs are going to be assigned to actual electrical pins on the FPGA.

Next we are making an "internal" wire, only usable within the module, and setting it to be equal to the result of input1 AND input2. Finally, we set the output to be equal to the value of this internal wire. We could have been more compact and just set the output to directly be the result of the AND, but in general for more complicated code using such wires greatly simplifies the readability of the code. The compiler will automatically compactify down the code for us, producing the same actual program for the FPGA in the end (e.g. no actual internal wire will be used in the FPGA either way).

Sequential Logic and Clocks

The previous code would be an example of "combinational" logic. The inputs are taken by the LUT and propagated into the assigned output as soon as possible. However, combinational logic becomes very unwieldy as programs grow larger and more complicated because you're never sure exactly how long the propagation of each LUT will take.

Another way of performing the AND logic would be using "sequential" logic:

```
output reg output1;
always @ (posedge in_clock) begin
  output1 <= input1 & input2;
end
```

The "begin" and "end" statements indicate a block of code that runs, here at each rising clock edge.

This output is only valid after the rising edge of the input "clock" signal, an additional input to the module. But we can at least be sure of the timings, and sophisticated tools can accurately predict the behavior and stability of large and complicated combinations of such sequential logic operations. How long it takes the output to be valid after the rising edge of the clock will determine the maximum clock speed at which the program can run. How fast a piece of combinational logic can run depends on the complexity of the logic (the longest propagation delay between dependent logic blocks). *Note*: all always blocks (and all combinatorial logic) run *simultaneously* (in parallel) in different logic blocks of the FPGA! The computational power of the FPGA is thus proportional to its number of logic blocks (and how fast they can run).

Getting a little fancier, you can also use a "register" to store the result, and also an "if" statement:

```
module and_gate_register (input1,input2,output1);
   input  input1;
   input  input2;
   output output1;

reg last_state=0; // register, will store the previous state
reg output1; // must also be declared a register

always @ (posedge in_clock) begin
   if (last_state==0) begin
     last_state<=1;
output1 <= 0;
   end
   else begin // last_state is 1 here
     last_state<=0;
output1 <= input1 & input2;
   end
end

endmodule
```

This will alternate each clock cycle between outputting the AND of input 1 and 2, and outputting 0.

Pin Assignments

Now that we have our logic defined, we need to interface our logic to the outside world. The FPGA talks to the outside world via *pins* that are on the outside edge of it, or for larger FPGAs on the bottom as solder balls of the ball-grid-array (BGA) chip package. FPGA software tools will usually have a graphical viewer of all the pins, and what those pins are assigned to. Each pin can be assigned as an input or output (or bi-directional), and given a name, which then connects it to that name in your design's HDL logic.

Pins are grouped in several *banks*, maybe two per side of the chip or so. They can each have their own IO voltage or *VCCIO*, by physically giving a certain voltage to that bank's voltage input pin(s). For instance, one bank might be used for talking to 3.3 V logic level chips and have a 3.3 V VCCIO, and another might have 1.8 V for talking to 1.8 V chips or using LVDS pairs (see below).

IO's can be single-ended, using just a single pin, or differential, using two pins, a + and − (or pos and neg). A common differential standard is *LVDS*, which is a $100\,\Omega$ differential standard. The IO is a 0 when the neg pin is higher in voltage than the pos pin, and 1 otherwise. When an IO is defined to be differential, the positive pin is assigned to a name, and then the negative pin is automatically assigned to the same IO.

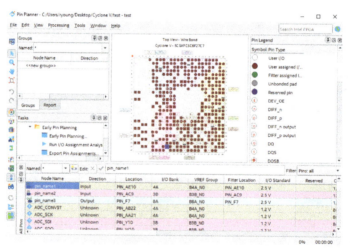

IP Blocks

It's always nice to not have to reinvent the wheel when writing programs. Just like a computer programming language lets you import *libraries* implementing others' code, FPGAs let you use others' modules, with well defined inputs and outputs. These can be included directly into your projects as source HDL code files.

But FPGA vendors also supply closed-source (even encrypted!) *intellectual property* (IP) blocks, which are like modules you can insert into your HDL and use. Often they make use of dedicated hardware resources in the particular FPGA, so are not just generic HDL. They are usually also parametric, in the sense that you can create them with various values that define what the IP block does. You could define the width of an input bus to be 4 bits, or 8 bits, for instance.

An important example of a hardware IP block is one that implements a *Phase-Locked-Loop* (PLL). It takes in an input clock of some particular frequency, and outputs several different output clocks, with frequencies that are some ratio (higher or lower) of the input frequency. The outputs can also have various phase relationships compared to the input clock. These are very useful, since often you need various clocks at various points in your design.

Boards and Programming

There are several FPGA manufacturers, Altera and Xilinx are currently the two big ones, but there's also Lattice, Efinix, and other upcoming brands. To get started with an FPGA one usually uses a *development board*, which has the FPGA on a PCB that supplies appropriate power voltages, supplies clock oscillators, provides convenient interfaces to the FPGA pins, etc. This is similar to how we used a complete Arduino board to use a microprocessor chip.

The FPGA company provides software for compiling the HDL, including IP blocks, and producing a *bitstream* file. That bitstream then gets programmed into the FPGA. This is usually done through *JTAG*, again using the provided software. It's usually also possible to write the compiled program into permanent flash memory on the board, so it loads the program automatically on bootup.

Lab: FPGA Lab

We'll be using the Step-Max10-02 board. This has an Intel (Altera) Max10 FPGA on it, so we use the Quartus FPGA software to program it: https://fpgasoftware.intel.com/18.1/?edition=lite&platform=windows.

Download the zip file of the Step Max10 code from github, then unzip it.

There is lots of example code. In Quartus, Open Project... and select `STEP-MAX10-master\examples\quick_start\lab1\led.qpf`.

To set pin assignments go to "Assignments -> Assignment Editor", and compare to the device schematics.

Make sure you are targeting the right FPGA. In "Assignment -> Device", select the 10M02SCM153C8G device.

To compile, press the "play" button, or "Processing -> Start Compilation".

To program, go to "Tools -> Programmer", and in that window press "Hardware Setup" and select USB Blaster by double-clicking on it, then Close. Press "Start" in the programmer window to program the FPGA. Note: it will revert to the default program stored on it when you power-cycle it! (There's an additional step that would be needed to overwrite the default program so it runs your code at power up.)

1. Use your FPGA development board to read external digital signals (buttons or switches), perform simple logic (like AND/OR) in verilog code, and output results via output pins (LEDs). Note that the FPGA sees a 1 when a button is not pressed (it is pulled high by a pullup resistor) and a 0 when pressed. Similarly the output LEDs are on when the FPGA is outputting a 0 and off for a 1.

2. Use a clock and sequential logic to make the output LEDs count up from 0 to 7, and then reset back to 0 and repeat. Or even try to use the 7-segment LEDs. It might help to use an array of registers:

```
reg [31:0] counter; // an array of 32 bits.
```

3. Input a signal to the FPGA from the signal generator to a GPIO pin on the board. Make sure that the voltage range of the signal matches that expected by the FPGA input (see in the Pin Planner), usually 0–3.3V. Look at the output from logic using that input (and the original input) on the oscilloscope. Measure the time delay between input and output, using combinatorial and sequential logic (using a clock signal).

Index